*The Private Eye*

Mary Burns

# The Private Eye: Observing Snow Geese

UBC PRESS / VANCOUVER

Printed in Canada on acid-free paper ∞

ISBN 0-7748-0575-7

**Canadian Cataloguing in Publication Data**

Burns, Mary, 1944-
  The private eye

  Includes bibliographical references and index.
  ISBN 0-7748-0575-7

  1. Snow goose. I. Title
QL696.A52B87 1996          598.4'1          C96-910494-4

UBC Press gratefully acknowledges the ongoing support to its publishing program from the Canada Council, the Province of British Columbia Cultural Services Branch, and the Department of Communications of the Government of Canada.

UBC Press
University of British Columbia
6344 Memorial Road
Vancouver, BC V6T 1Z2
(604) 822-3259
Fax: 1-800-668-0821
E-mail: orders@ubcpress.ubc.ca
http://www.ubcpress.ubc.ca

*This book is dedicated to David Hughes Robertson*
*with thanks for his faith and his always generous support,*
*and to the memory of Glen Smith,*
*wildlife biologist, bird artist, gentle man.*

# Contents

# Maps and Illustrations

# Introduction

*4 December 1992.* Just below zero on the Celsius scale, about thirty degrees Fahrenheit. Brilliantly sunny. I arrive at the Reifel Sanctuary just on noon, high tide today, when the great flocks of lesser snow geese that spend much of the winter here customarily fly in from the shore of Georgia Strait to converge on the agricultural fields that spread over Westham Island, one of the few rural precincts remaining in the Fraser River estuary, some twenty kilometres south of Vancouver, British Columbia.

It is high tide, I'm certain, but I see no geese at all in the open rectangle of chewed-down timothy and rye where I usually find them. A great blue heron statuesquely patrols the slough that trickles alongside the road leading into the sanctuary; some coots and wigeons and mallards paddle sepia water while others squat on the ice lenses that scallop the brown mud shore. But there isn't a snow goose in sight.

Then I hear something, a broadcast turned low, gradually growing in volume as the first squad begins sailing in from the foreshore; less honky than the calls of Canada geese, more like boy sopranos stretching for their highest notes. Wavy lines of them advance unevenly, like a group of marathoners taking off in consecutive heats. Their wing tips rise, putting brakes on flight as the birds reach the field's end, curl back, settle on the earth. They land in groups of three to four, sometimes larger, and groups nearby flutter up, fuss with finding the right spot. Their restlessness seems an attempt to make room for the rest of the flock, for choruses in the distance announce a new wave, then another. This continues for ten to fifteen minutes until the field I've been watching is white with the presence of several thousand snows.

The few Canada geese that were feeding here have fled. The starlings and blackbirds and chickadees that fret the roadside alders and berry bushes with their twittering, their songs, hush in the presence of this legion, or perhaps are only inaudible, secondary now.

All the while I've been watching, car after car has been passing by my lookout spot on the road edge. Many cars, many more cars than I would have expected to encounter midday, Friday, the 4th of December. The snow geese are a spectacular sight to be sure and it's gratifying to know so many appreciate it, yet the glimpses I've caught of people behind the rolled up windows of their vehicles indicate that this ever growing group is not dressed for hiking or birding. The geese continue to sail in from the foreshore, and the business-suited men and women drive in the opposite direction, to the parking lot of the Reifel Sanctuary. Unable to contain my curiosity, I climb into my car and follow them down the road to where the normally commodious patch of gravel is so crammed a few people are actually directing traffic.

'What's going on?' I shout to the man who is rotating his arm in a signal to drive on, past the house of the sanctuary resident manager, back to where I entered the lot.

'The prime minister's coming.' He sounds a little suspicious, as if he thinks I am trying to crash this group of welcomers, party faithful, bureaucrats, local businesspeople, whoever they all are.

'Oh,' I say, comprehending finally. 'I thought everyone came to see the geese.'

He nods, too preoccupied to smile. The people assembled stand in groups, talking, waiting, and the snow geese keep coming, in the undulating threads that distinguish their kind. I think of the lines that embroider the shore after the tide recedes. What would happen if the tidelines could rise, develop flight, drag the sea's influence farther onto land? But it's no time for fancy. I'm holding up traffic.

Back at my observation post alongside the field I notice dark-suited men talking on cell phones. I notice a black limousine speeding along the normally sleepy lane. It must be him. But it isn't. Not yet. Only a car of security men, preparing the way. Several minutes tick by before I turn my eyes to the road again as a second limousine approaches, this one speeding too. Through the smoked glass of the rear window I catch a glimpse of the familiar deck of steel-coloured hair, the pressed collar, the overcoat, the man himself.

THIS IS a true story. On 4 December 1992, Brian Mulroney, then prime minister of Canada, visited the George C. Reifel Migratory Bird Sanctuary to sign documents that ratified Canada's commitment to the international initiatives on biodiversity and climate change that emerged from the 1992 Earth Summit in Rio de Janeiro. The press release I later obtained from Canada's Ministry of the Environment did not mention who attended the signing ceremony that day, or how many exactly were there.

So as I recall and write the story, from notes I jotted down at the time, embellishments creep in. Were all the welcomers dressed in business suits, or do I suppose they were because I imagine they were Conservative party faithful or Ministry of Environment bureaucrats? Have I based my assumptions on stereotypes? And the geese, did they really sweep in that directly over the heads of the people gathered there in the parking lot? Or do I just remember it that way because the image makes such a perfect cartoon: the stretched goose neck, flapping wings stirring the air over people concerned only, it seemed to me, with the appearance of fealty to their leader.

As a writer of fiction I understand that the point of view I choose for a short story or a novel determines the tale that will unfold. Much as that seems a discovery each time I struggle with the beginning of a new work, in the real world it is accepted as commonplace, I believe, that reality changes with the viewer. At the same time we operate as though there is an objective reality. 'Just the facts, ma'am,' said the cop on the classic TV show *Dragnet*. 'I want the truth,' we tell our children. Our words suggest we believe that we all see the same things the same way, even though we may view them at different times, from different angles, at different stages of our life, and for different purposes. A politician and a birder would have different accounts to tell of the early afternoon at the Reifel Migratory Bird Sanctuary on 4 December 1992, and both would be true according to the teller.

I began watching the snow geese on Westham Island for personal reasons: I wanted to become more observant generally, and to learn more about the natural world. I chose the Wrangel Island, Russia, population of lesser snow geese to observe because, for the part of their life history that takes place here in the south, on the Fraser and Skagit River Deltas, the snow geese provide a spectacle that's irresistible to anyone who has any romanticism at all in his or her soul. I had the familiar human urge expressed in the phrase, 'Hey, did you see that?' – to share the thrill. The very name snow goose conjures up images of the cold and distant barrens most of us have never seen, will never see. Or we think of Paul Gallico's lyrical short novel *The Snow Goose*, about the terrors of war and the loneliness of a man and the endearingly mysterious behaviours of individuals who have lost their flock, be they goose or human.

I liked the name, snow goose, and I liked the sight of them. Though I know better than to use it, the word breathtaking actually describes exactly what happened the first time I walked on the marshes of the Fraser Delta and was surprised, then stopped breathless for a moment, by the sudden rising up of tens of thousands of snow geese at once, the airy tumult of their madly beating black-tipped wings, the high soprano bark of their calls. I described them to someone as poetic, the way they stretch out across the sky like the broken lines of verse.

And they were accessible. On days when I had no classes to teach and my daughter was in school and the tide was up and the fields offered plenty of forage, I could park or walk along various Westham Island roads mere feet from the geese. I could listen in on their incessant chatter, look and look and discover what I learned just by looking. Then look again, an informed observer as time went on and the various biologists, artists, naturalists I talked to alerted me to aspects of the geese I might not have discovered in the short bursts of time I spent watching them myself.

I learned about the geese, and as I did I learned something about the people who were telling me what I wanted to know. It seemed I had a perfect opportunity to investigate the complexity of the human view of the natural world, to try to see that world – at least the fragment of it I chose as my focus – through the eyes of the scientist, the naturalist, the artist, the farmer, the hunter – and the Native people whose close-up view of nature evolved from their direct dependence on it. This book, then, is as much about the individuality of perception as it is about snow geese.

Numerous people generously assisted me in my explorations. I am particularly indebted to Fred Cooke, senior chair in wildlife ecology at Simon Fraser University, his associates Evan Cooch and Barbara Ganter, and to Sean Boyd, Canadian Wildlife Service, Delta, who patiently answered many phone calls and supplied much information on an ongoing basis. Also to Graham Cooch of New Mexico State University, Las Cruces; Mike Davison, Skagit area wildlife biologist for the Washington Department of Fish and Wildlife; Charles Hunt, US Fish and Wildlife Service, Bethel, Alaska, for his enjoyable and informative letters; Cynthia Wentworth of the US Fish and Wildlife Service; technicians Saul Schneider and Barbara Pohl; Mike Samuel of the Wildlife Health Center in Madison, Wisconsin; anthropologists Julie Cruikshank, David Ellis, Robert J. Wolfe, Bruce Millar, and Ann Fienup-Riordan; Robert Bateman, who enthusiastically engaged in some spirited conversation with me about his perception of nature; John Ireland and Varri Johnson of the Reifel Sanctuary; Maynard Axelson, who shared experiences and photographs; Robert Husband and Henry Parker, who took me hunting; Dave Smith of the Canadian Wildlife Service, and BC Ministry of Environment staff who supplied me with figures and information; Julie Cruikshank, Sean Boyd, and Dr. Helen Hossie, who provided invaluable reflection on early drafts of some chapters; my sister Mike, who shared my enthusiasm; UBC Press senior editor Jean Wilson, whose faith in the project provided the necessary encouragement; and my daughters Elisabeth and Annie for their interest, their companionship, and their help.

I am also grateful to Edmund Blair Bolles for the insights and information in his book, *A Second Way of Knowing: The Riddle of Human Perception*, which helped guide my thinking and writing in Chapter 2. *The Private Eye* quotes several sources, and an honest attempt has been made to secure permission to reproduce small portions of text from those quoted most extensively, as follows:

From pp. 2-3 of E.G. Cooch, L.C. Newell, R.F. Rockwell, and F. Cooke, 1990, Queen's University Tundra Biology Station Lesser Snow Goose Project Summer Guide. Unpublished

From p. 186 of Fred Cooke, Robert Rockwell, and David Lank, 1995, *The Snow Geese of La Pérouse Bay*. New York: Oxford University Press

From p. 33 of Edward Dobb, Without Earth There Is No Heaven, *Harper's Magazine*, February 1995

From pp. 72, 73, 168 of Ann Fienup-Riordan, 1990, *Eskimo Essays*. New Brunswick, NJ: Rutgers University Press

From *The Snow Goose* by Paul Gallico, Copyright 1940 by The Curtis Publishing Company and renewed 1968 by Paul Gallico. Reprinted by permission of Alfred A. Knopf Inc.

From R.H. Smythe, 1975, *Vision in the Animal World*. London: Macmillan

From p. 37 of E.V. Syroechkovsky and Fred Cooke, n.d., A Comparison of the Nesting Ecology of the Lesser Snow Geese of La Pérouse Bay, Manitoba, and of Wrangel Island, Chukotka, USSR. Unpublished

From H.R. Thornton, *Among the Eskimos of Wales, Alaska, 1890-93*, 1931. Reprinted by permission of Johns Hopkins University Press

From pp. 18-19 of *How Birds Fly and Other Marvels of the Animal World*, copyright © 1992 by Michael Friedman Publishing Group Inc. Reproduced with the permission of Doubleday Canada Limited and Michael Friedman Publishing Group Inc., 15 West 26th Street, New York, New York 10010

# The Private Eye

# The Flight South

THE NORTH AMERICAN CONTINENT curls like a canehead north and west to within figurative hand-clasping distance of the Russian mainland, which lies across Bering Strait. Less than eighty-five kilometres, fifty-three miles, separate Siberia from Alaska. A few degrees south of this point, below Norton Sound, the Yukon and Kuskokwim Rivers separate into an intricate earth-toned mosaic of streams and sandbars, sloughs and lakes and bays.

In mid-September, as waterfowl migrating south from their various arctic nesting grounds join local birds to feed on the sedges and grasses that tuft the rich delta soil, the gauzy skies above the rivers' mouths form the backdrop for a virtual ballet of avian diversity. There are birds, particularly waterfowl, everywhere: spectacled, Steller's, and common eiders; mallards, pintails, wigeons, greater scaup, and scoters; tundra swans, lesser sandhill cranes, emperor geese, Pacific white-fronted geese, black brant, and several species of shorebirds and songbirds all present in profuse numbers. But the largest flocks in the delta, especially around the village of Kotlik at the north mouth of the Yukon River, are those of the lesser snow geese, which have been feeding here since early September when they flew across the Bering Sea from the east coast of the Russian mainland, the first stop as they migrate south from their nesting colony on Wrangel Island, Russia.

The snow goose (*Anser caerulescens caerulescens*) has plenty of features that distinguish it from others of the genus *Anser*, most noticeably smoke-black wing tips that contrast handsomely with the white feathers that cover the rest of its body; a smooth, curved head; round dark brown eyes that appear startled, as eyes seem to be when not bounded by eyebrows; dusky coral feet and legs; and a saw-tooth edged, black-lipped bill that recalls an actor playing a gangster talking out of the side of his mouth: Jimmy Cagney, maybe, or Edward G. Robinson. Biologists describe it as a grinning patch.

And then there is their call: Roger Tory Peterson characterizes it as 'a loud, nasal, double-noted houck-houck in chorus'; other ornithologists as 'a high pitched, *ou, ou*, somewhat similar to the bark of a fox terrier.' Still another bird book says to listen for a short muffled *hau-hau, hau-hau*, like the tundra swan's. Long before taxonomists gave the snow goose a proper scientific name, Eskimo people heard the sound 'kanguq, kanguq!' and thereafter called the wild white geese that flew over each fall and each spring by that name.

The first flocks to land on the delta were small sub-groups of the main flock, consisting of yearlings and those adults that did not successfully breed in the late spring. But within a week or so, the dots of white that began appearing about the second week of September have multiplied and merged in puddles, entire lakes of white, and the colour mixture has become more complex. Mixed in with the white of the yearlings and adults are greyish young, the new generation. These almost full-grown goslings hatched in late June, and learned to fly mere weeks ago, just after their parents' flight feathers grew back. Their inexperience makes them easy targets for the Yup'ik hunters who motor out the estuary in boats and wait with shotguns ready for the big flock to fly up. The safety-in-numbers factor is one reason why snow geese stay in family groups that join with other family groups on the feeding grounds.

Because their food is low in nutrient value, and, unlike cows, they

Snow goose holding grass in its serrated bill

have no rumen to break down the cellulose they ingest, they have to eat almost all day long. And the females and young in particular greatly need fat and protein to fuel them for the flight south, for during the twenty-two days she sat on her eggs, the mother snow goose rarely left the nest to feed. They use their bill to tear the roots and grasses from the marsh, then hold the plants with the serrated edge of the grinning patch so that their powerful, toughened tongue, the edges of which are also serrated, can cut through the roots. They do this all day long and into the night, resting for short periods while a family member keeps an eye out for predators, chattering in between bites, but mostly feeding, feeding.

They stay well out on the marsh, out of the range of the hunters from Kotlik and other delta communities whose bird harvests help fill their families' larders. Compared to their annual take of mallard, pintail, and scaup, the Yup'ik hunters have little impact on the size of the flock migrating this fall. They, and other, mostly white, hunters along the northwest coast of the continent, in Alaska, British Columbia, Washington State, and in the fields of California, will reduce the population by only about 10 per cent in a given year.

It is the conditions on Wrangel Island itself, the sudden unpredictable freezes, the everlasting snow, and the state of the predator-prey balance that determine how many goslings will survive to eventually reproduce themselves. The 1994 breeding season was a tough one. While in normal years about 20,000 to 30,000 pairs of snow geese nest on the Wrangel colony, in 1994 only about 12,000 pairs even attempted to nest. By the time the snow started to melt in early June, all the others had decided it was too late and too cold. They didn't even try. The ones that did nest were about two to three weeks later than normal because it was so cold in mid- to late May. For two to three days at the beginning of June the ice and snow melted rapidly and the geese began to lay their eggs. Then, when the female had started to incubate them, snow started to fall again and it snowed every single day in June. Of the 12,000 nests, only 2,000 produced eggs. And some of the goslings were taken by arctic foxes shortly after hatch. Still more were lost in the perilous first stage of migration, from Wrangel to the Chukotka Peninsula, when inclement weather tested the endurance of the new fledglings and many failed.

So in 1994 only a few of the heads dipping into the mud at the mouth of the Yukon River were the brownish-grey of that year's young. And because they were smaller and weaker, each next step in migration would threaten even their survival. In 1994, as in 1989, there was a chance that there would be no new generation of Wrangel snow geese.

It's one of those years that demonstrates the fragility of this otherwise healthy species, the complete opposite of 1993, when the goslings made up 40 per cent of the population. In good years such as that, the young males are usually about 2.5 centimetres shorter than their male parent, who averages about 74 centimetres, and weigh only about 200 grams less by the time the flock leaves the Yukon Delta. From an average of 100 grams at hatch, the goslings, male and female, grow to 1,200 grams in six or seven weeks. They add to their bulk on the delta, as do the adults, the male acquiring 14 per cent of his overall weight and the female 17 per cent.

But nights are getting longer and colder. The sun does not rise until eight in the morning. Bright red rosehips decorate the spiny branches of the wild rose bushes upriver, and the leaves that still cling to the berry bushes on the delta are yellowy-brown, brittle. The first good wind will tear them off at the stem. Farther upstream the banks of the Yukon and its tributaries reek of spawned out silver and pink salmon.

The big flocks that have been feeding for three weeks to the west of Kotlik head slightly up the Yukon River now to just northeast of Scammon Bay, then onto the east side of the Askinuk Mountains where they feed and rest for a few days.

By the time they arrive at Baird Inlet, which is their last stop to feed and rest, most of the lakes, rivers, and sloughs in the Yukon-Kuskokwim Delta are frozen. The few young snow geese and their parents, and the yearlings and the many failed breeders register the change in the earth's axis, sunlight diminishing every day, and respond with a program imprinted on their genes. Perhaps it is one of the mature failed breeders, or two, a pair, that react first and jump into the air as snow geese do, being one of the species that do not need a running start. The families with which these two were feeding rise seconds after. Then other groups join and follow, continuing the journey that began on a northeast Siberian island

and will end, for more than half this population, 4,850 kilometres south on the Fraser River Delta just outside Vancouver and a little farther south, across another national border, on the Skagit River Delta in northern Washington State. The rest of the flock will continue via different routes a farther 1,220 kilometres to California, to spend the winter on a refuge near Sacramento.

The fact that snow geese mate for life and spend their lives in family groups has inspired poets and essayists to consider them romantic creatures, wiser in some ways than humans. It is an equally homey sort of notion to imagine that the unusually large flocks in which they migrate – 100 to 1,000 or more – are made up of relatives – cousins, aunts, uncles, grandparents. However no one knows for sure if this is the case. What biologists do suspect is that families that fly together know each other from the breeding grounds. They stop at the same places on their way south because the marshes that provide good grazing are like the proverbial needle in the haystack, postage stamps stuck on far-between spots on the rambling Pacific coast. The birds that have survived are those that know these places, and they've shared their knowledge with their cohorts, and their descendants, who copy them.

A single family within the flock may contain the parents, the juveniles, and the young adults who hatched last year and have not yet found a mate themselves, for young snow geese stay with their parents for the entire first year and fly back to the breeding grounds with them. Because they do not breed until their second at the earliest, most commonly their third or even their fourth year, young geese stay on the fringes of the breeding colony and head for the moulting ground early, to shed their flight feathers and feed with other young geese. Some then meet up with their parents, and their siblings, and travel back south with them.

The goslings that made it to the Yukon Delta have gained enough weight that their chances of going all the way, to the wintering grounds on the Fraser and Skagit Deltas, and the rice fields in California, are greatly improved. But the journey is not without danger. For while late September along the Alaskan and British Columbian coast can be the best of times, the bright, crisp Indian summer that is sweeter because we know it cannot last, the severe weather the geese escape by migrating can

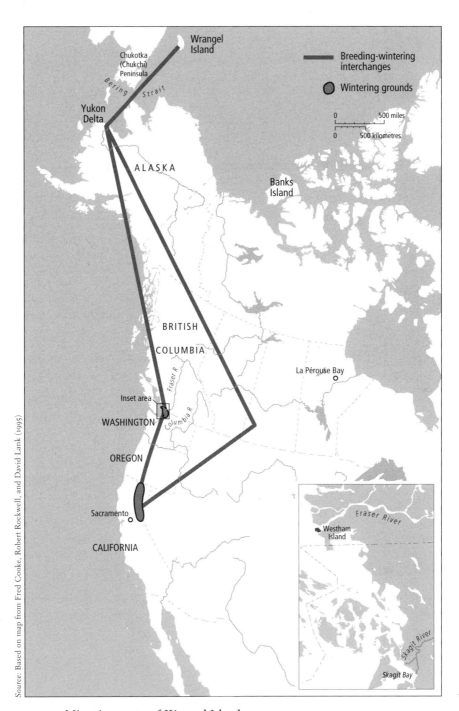

Source: Based on map from Fred Cooke, Robert Rockwell, and David Lank (1995)

Migration routes of Wrangel Island snow geese

follow them. A storm front moving south from the Arctic can batter the coast with currents too violent for the young to withstand. The flock can splinter apart, families, individuals lose their way.

But snow geese are the noisiest of all waterfowl. You can hear the calls of a large flock about to take flight or to land a mile or more away. And their loud voices help them migrate successfully, for migrating birds stay on course by honking, croaking, trilling, or whatever it is they do, and listening for responses from their own kind.

'In this way,' writes Roger Caras, in his book *The Endless Migrations*, 'each bird has scores, hundreds, at times hundreds of thousands of similar eyes, ears and genetically coded migratory patterns to rely on, not just its own limited physical package.' Hearing the sound of the flock, stragglers can position themselves and find a way back to it.

They usually start their migration after sunset but continue flying night and day, at speeds of about sixty to eighty kilometres per hour. They often fly with a tailwind to boost speed and reduce energy needs. Though average flight altitude is 610 to 914 metres, airplane pilots have reported seeing them as high as 6,096 metres and it is not uncommon for a flock to reach altitudes ranging from 1,500 to 3,000 metres. They may beat up to higher elevations to escape a storm, or drop below high clouds for the same reason.

Their long narrow wings taper to a point and tend to be swept backwards. The distinctive black primary feathers are attached to hard bones that they drive through the air with large flight muscles attached to the ridge along the breastbone, called the keel. The secondary feathers along the wing are responsible for lift. Each feather functions independently so that the goose can alter the shape of the wing during flight. The geese hold their wings at a slight angle to an air current: the air flows faster over the upper surface than the lower, and the loss in pressure above the wing causes lift. At the same time, resistance to moving air tends to drag the wing backwards. By adjusting lift and drag to equal its weight, the bird is able to glide.

The powerful flapping flight of the snow goose during migration is achieved by the inner wing: while the secondary feathers give lift, the hard section, the primaries, pull the body forward. During the power

Snow geese take flight.

stroke the primaries are linked together to form a near perfect aerofoil, giving maximum thrust and maximum drag.

Pushing steadily south in this manner, the geese reach the Stikine River Delta in northern British Columbia about three to four weeks after they have left Wrangel Island. The Stikine empties into the Pacific at the top of the Alaskan panhandle, across from a Wrangell with an extra *l*, this one located in US territory: Wrangell, Alaska. Somewhere around here many of the snow geese destined to winter in California wheel inland and fly across the coast mountains to continue their journey south through Alberta to Summer Lake in Oregon and finally Tule Lake in California's central valley. Eighty per cent of Wrangel nesters used to settle in California for the winter, but in the last twenty-five years the percentage has dropped to half that amount.

So now the flocks spread out in skeins high above the coast consist of about 40,000 to 50,000 birds or more, including the grey-feathered juveniles making their first trip. From the Stikine the journey is straight south, above scores of inlets and bays, estuaries of wild rivers such as the Skeena and the Khutzeymateen, great salmon rivers where bald eagles perch expectantly on snags, over the puzzle of land and water that leads to the passage separating the mainland of British Columbia from Vancouver Island and the open water of the North Pacific.

I imagine an enormous composition in black and white as the conifers darken in the faltering autumn light, the alder and broadleaf maple, the cottonwood and willow lose their leaves, the first snow dusts mountaintops and hillsides, gleaming jet and ivory orca move in and out of the water, blowing, and migrating birds of dozens of different species sketch the sky with scattered dots and spirals and arrows and undulating lines.

But it is impossible, of course, to see into the goose brain, to see what snow geese see. Perhaps they don't even look down, but continue as if driven to their final destination. Geese, like most other birds, have retained their third eyelid and during the long migration flight they keep it down, a kind of veil to protect the sensitive retina. Who knows if they register the black and whiteness of the world below, if they sense the opera of which they might be the chorus, calling?

# 2

# The Eye of the Beholder

I PERSONALLY saw nothing of the picture I just presented. I pieced together bits of visions provided me by various biologists and naturalists, with imagination for the glue. Scanning my words I notice I have already compared snow geese to marathoners, to opera singers, to Jimmy Cagney. I realize I am applying a human perspective, something that makes me nervous for its reductionism, its anthropomorphism. I never intended this to be a biography of goosey gander and his friends. Yet humans are bound to perceive as humans, to take the human view.

The mechanical process of seeing is often explained by describing how a camera works. Light rays strike an object such as a snow goose: some rays are absorbed but the portion of light that is reflected passes through the lens of our eyes, and falls on the retina inside the eyeball, which acts somewhat like the film inside a camera. The retina is made up of many separate cells: each registers part of the object. These fragments are sent to the brain along a web of optic nerves, and the cells that make up the visual cortex of the brain, which has stored information previously perceived, 'develop' the separate images into a total picture.

Because we have two eyes placed where they are in our heads we have binocular vision: both eyes look in the same direction and their range overlaps. The shape of our eyeballs determines that normally functioning human eyes have the ability to perceive depth and breadth. The other components of vision are colour, form, and motion. When we look at a scene, such as a field of grazing snow geese, the eye acting in cooperation with the brain translates the stimuli conveyed by the light rays into a visual image. Fully a third of the brain is devoted to processing what we see.

Seeing. Noticing. Glimpsing, glancing, gazing. The progression from observing to perceiving to recording to responding. Love comes in through the eye, said the Irish playwright Sean O'Casey. But does everything we know and feel enter through the same doors, and if it does, what made me select the snow geese from all the things there were to view that December day at the Reifel Sanctuary, while others of my kind focused on their political chief?

Sir Francis Crick, co-discoverer of how DNA reproduces itself, wrote that perception is science's greatest mystery. While it is relatively easy to describe the physical process of seeing, scientists are still trying to explain how mechanistic inputs lead to awareness and understanding. Common sense tells us that we are not aware of every detail a scene presents. Though I saw – I noticed, *perceived* – both the geese and the visiting prime minister, the picture before me everywhere I turned was as rich with detail as life itself. I think of my daughter's activity book – Can You Find the Hidden Picture? Did I notice the voles zipping through the grass, the hawk waiting in the pear tree, the heron camouflaged by the sloughside brush?

In *Vision in the Animal World*, R.H. Smythe writes: 'Eyes which become too efficient might raise problems and cause their owner a deal of embarrassment. Man has overcome this possibility by seeing a lot but observing very little. If he walks the length of a street, engaged with his own thoughts, he will probably see several people ... But it might happen that one person he met was of considerable interest and the impression he had registered of that person might enable him to describe him, or her, in great detail. His eye is merely an optical instrument. It is not the eye that sees but the brain behind it. The eye merely observes and the brain behind it decides whether or not to develop the picture.'

According to Gestalt psychology, figures are the things we pay attention to; ground is the assembly of sensations we ignore. Perception begins with the sorting of figure from ground. We determine the meaning of what we see by registering its identity – what the figure is, and how it differs from others of its kind, its individuality. Our own experience – memories of what we have seen and undergone before – controls the process of selecting details.

Yet in our everyday life we subscribe to the notion that there can be impartial witnesses, whether as reporters describing events for the evening news, or in court. Psychologist Robert Buckhout, in *Scientific American*, says that those who depend on eyewitness testimony in court, depend on an outdated nineteenth-century view of humans as perceivers, a view that asserted a parallel between the mechanisms of the physical world and those of the brain. 'Human perception is a more complex information-processing mechanism. So is memory ... Human perception and memory function by being selective and constructive. Perception and memory are decision-making processes affected by the totality of a person's abilities, background, attitudes, motives and beliefs, by the environment and by the way his recollection is eventually tested.'

Given this and other studies of perception that highlight the singularity of each view, it's remarkable that we operate on common principles at all. Language helps. When people in the same culture agree on the common characteristics of an object, they give it a name. The business of taxonomy is all about classifying living and extinct organisms into a hierarchy of groups that have some logical relationship to one another. By speaking the same language we more or less acknowledge that we generally view the world the same way. That this is a wildly optimistic simplification, however, becomes obvious with a quick survey of two dictionaries. According to *Webster's New Collegiate*, for example, *see* means '1. to perceive by the eye'; but also '2. to have experience of, 3. to form a mental picture of or perceive the meaning of, or to imagine as a possibility.' *The Concise Oxford* defines *see* as follows: '1. to have or exercise the power of discerning objects with the eyes.'

Both definitions started me on a familiar, baffling quest for clarity, something I think of as the dictionary game, as I leafed through the *p*'s for perceive, the *d*'s for discern. The consistent if minor variations in definitions demonstrated that individuality of perception extends even to the writers of dictionaries. While the differences might seem trifling, think about the impressions evoked by the names orca and killer whale, which both refer to the same marine mammal.

That language reflects the individuality of perception is also obvious from these two descriptions of the vegetation at La Pérouse Bay, Manitoba, site of Dr. Fred Cooke's twenty-five-year-plus study of lesser snow geese. The first is from the Russian snow goose biologist E.V. Syroechkovsky. 'In the region of La Pérouse Bay ... shrubs and scrubby vegetation predominate and the shrubs consist of several species of willow and birch which reach 1.5 to 2.5 metres in many places, forming real thickets, difficult to penetrate. The height of the bushes increases in the direction from the shore of La Pérouse Bay to inland.' The artist Robert Bateman had this recollection of the same place: 'I was in this bounteous world, eider ducks and oldsquaws and ptarmigan, and beautiful weather like this and sloshing in our hip waders through this champagne coloured shallow water, stepping on these little knolls that were just like bouquets of flowers and little bonsai trees.'

In his book, *A Second Way of Knowing: The Riddle of Human Perception*, Edmund Blair Bolles traces the history of scientific thought about perception, concentrating on the division between 'physicalists' and 'humanists.' Bolles spent considerable time with computer scientists who are trying to invent a computer that will perceive as humans do, and keep running up against the limitations of the mechanical model. 'At the heart of the challenge in getting machines to perceive things lies a basic contradiction,' Bolles writes. 'Machines require standardized input, but we live in an unstandardized world.'

Apparent proof that perception is subjective rather than objective came in the 1980s when colour vision theorists discovered that while the retina receives mechanical information – wavelengths, in this case – it does not compute a numerical wavelength but transmits a qualitative description to the brain. 'The nervous system's change from physical to subjective descriptions of the input begins at the sensory receptors,' Bolles asserts.

That doesn't mean the diehard 'physicalists' have given up. They still refuse to accept that something beyond physical sensation exists in perception. They are convinced that scientists will eventually be able to break down the mechanics of perception into parts fine enough to duplicate. It is only a matter of time, and more funding for research. 'Humanists,' as Bolles calls them, insist this is impossible: the subjectivity that comes into play when eyes see colour, and presumably everything else, proves it is impossible. Generalizing in a way that seems to give the lie to the results of his own research, Bolles puts most scientists in the 'physicalist' category, artists on the 'humanist' side. Scientists observe to turn away and generalize, he says; artists to seize and use reality. And he quotes an essay on the divided self by the late novelist and philosopher Walker Percy to bolster his point: 'There is a difference between being-in-the-world of the scientist and being-in-the-world of the layman.'

Both men clearly believe that this dichotomy in human thought about perception is big trouble. Walker Percy calls it (and the essay in which he addresses it) 'The Fateful Rift: The San Andreas Fault in the Modern Mind.' While undoubtedly excited about what scientists may yet discover, Bolles says: 'The holy grail of a machine that understands the particular meaning of a particular situation, however, is a dream. People who seek that grail are hunting for a road that leads directly from objectivity across the uncrossable divide into subjectivity.' Bolles thinks there is an unbridgeable principle in psychology just as there is an uncertainty principle in physics, referring to Heisenberg's discovery that each particle has a wave. He rues the fact that 'physicalists' remain unconvinced, and that people in general tend to recoil from descriptions such as subjective and opinionated. 'We all know that we are more subjective and less predictable than machines but we consider these characteristics to be weaknesses.' He concludes by urging us to 'trust the mystery' of perception.

Walker Percy suggests we use language to bridge the gap, saying, in a way that requires some nimble synaptic leaps to comprehend, that there can be no gap if we identify it as such; it's enough to call it a gap and accept that this is part of the way we 'are' in the world. He has taken a different route to an end by also asking us to trust the mystery that exists between mind and matter.

The differences in how people 'see' things is the gist of controversy. And attempts to create public policy on controversial issues, such as the management of wildlife in populous areas, must take the subjective nature of perception into account. To paraphrase Walker Percy, there is a difference between being-in-the-world of the city person and being-in-the-world of the country person, otherwise known as the urban-rural split. The problem is that neither generalization captures the unique histories of the individuals who form these constituencies. Governments tend to deal with interest groups. Because there is power, if not always precision, in numbers people tend to join with others who share similar feelings to try to change public policy. In North America, the public hearing, the town meeting, where people gather to express their views to the powers-that-be, is a long-standing tradition.

Thus in March 1993, the Canadian and British Columbia governments, through Environment Canada and the British Columbia Ministry of Environment, placed an advertisement in the *Vancouver Sun* inviting people to bring their questions, concerns, and suggestions to a public meeting dealing with urban encroachment on wildlife habitat, hunting, and the management of brant, Canada geese, and snow geese on the Lower Mainland, a topographically vague description of what is in fact the delta of the Fraser River.

**23 March 1993.** The Scottish Cultural Centre, Vancouver. By seven o'clock, the large auditorium is already busy with men and women wandering around, examining the various displays, maps, photographs, and pouring over some of the many 'Fact Sheets' the two levels of government have prepared. Straight-backed chairs handsomely upholstered in dark blue fabric are lined up in rows facing the stage, which is framed by a tartan curtain, the Cameron plaid. China cups and saucers in the blue willow pattern – a major improvement over the usual styrofoam – are arranged next to the coffee urn. So it's busy and congenial, and a visually pleasing place to be, except this is not a smoke-free building, and the air has a tight, stale edge to it, as cigarette exhalations drift in from the foyer.

The moderator, Dick Young, of the Delta Environmental Advisory Committee, a tall balding man with rimless glasses and a full wiry beard and moustache, introduces the panel of experts assigned to speak on present public policy and answer questions. From the vantage point of these five men and one woman it might seem as if the audience is largely divided according to right and left, the hunting groups on the right, the naturalists and humane societies on the left. Of course there are exceptions to this arrangement, independents such as myself who, not knowing any of these people personally or what they represent, sit among a group of what turns out to be hunters. About two hundred people altogether, I estimate, more men than women.

On some issues, there is general agreement. No one wants developers to get their hands on Burns Bog, a 4,000-hectare deposit of sphagnum moss where plants such as Labrador tea, cloudberry, sundew, and reindeer moss grow at their southern limit, providing habitat for 150 species of animals. Fearing more subdivisions and golf courses in an area that once supported large populations of sandhill cranes, Eliza Olsen of the Burns Bog Conservation Society pleads with the government to buy the bog. A speaker from a local fish and game club supports the idea. In fact there seems to be a natural alliance between the farmers and hunters and naturalists. In this case, their antagonism is directed towards the government. A clear sense of distrust emerges, with good reason. The problem is money. 'We can't buy it to save,' says the provincial government representative. 'It's too much land.'

'That's exactly why it should be saved,' someone in the audience argues back.

The topic is urban encroachment, creeping citification, golf courses attracting housing developments whose occupants want shopping malls nearby, whose children need schools. And the reason the city of Vancouver is spilling outwards, onto land that once supported enormous numbers of wild creatures, is that people, like wildlife, find southwestern British Columbia congenial. It wouldn't be so much of a problem, a few people imply, if the government restricted immigration. The territorial tone of the argument begins to emerge.

Barry Leach, naturalist and the author of *Waterfowl on a Pacific*

*Estuary*, cannot wait to complain about what he sees as government neglect. An articulate, slender, patrician-looking man in his mid-sixties, long active on behalf of birds, he's angry that no one seems worried about how population growth is threatening to decimate natural areas around Boundary Bay, an enclave of mudflats and eel grass that is home to perhaps the largest congregations of wintering waterfowl in Canada. Ethnic groups are going to scrape all life off the beaches, he claims, referring, I think, to crab fishers who have emigrated from southeast Asia. He speaks of experiences in his native England, drawing the scorn of some men behind me, who say, when he mentions the Thames, in England, and Germany's Weser, 'Where's that?'

The hunters are, as a group, burly rather than slight, bearded or moustached rather than clean shaven, and many wear billed caps. One tall, white-haired man among them heads for the microphone in the centre aisle to continue the rant against immigrants. In response to Leach, whose English accent has made him a sitting duck, so to speak, this person starts by saying: 'I didn't emigrate from anywhere. I was born right here.' After that statement the speakers seem to feel they must validate themselves by reporting how long they have been resident in the delta, in British Columbia, in Canada, before they say their piece. It's as if they are saying that long habitation promotes clearer vision, subconsciously trying to remind their fellows of the difference between glancing and gazing.

It's clear that these hunters feel attacked. They say that the 8,000 people who attend the snow goose festival each year do more harm than the hunters to waterfowl. The group behind me joke, in low voices, about being treated as child molesters.

As the evening wears on and speakers from the left, right, and middle of the room, and the same spectrum of sentiments, voice their concerns, the expressions on the faces of the panel members tell me that they know many of these people and what they are about. On their part, I hear audience members referring to the Fact Sheets that were handed out as 'more of the same old shit.' Both the panel and the audience seem to feel under appreciated. So what has been accomplished? The hunters ask for longer days – an hour after sunset rather than half an hour, a longer brant season now that they've been put off the beaches. The naturalists ask for more protected lands. The farmers want monetary compensation for the crops all the protected waterfowl consume.

The responses from the representatives of the various government agencies are received with scepticism. The meeting may have let off some of the pressure that has built up within groups of people who feature different scenarios for the wildlife-human interface. Or a small step might have been taken towards finding out what is the best thing for all humans and birds concerned, but has anyone actually gained insight here? Real penetration into another's view?

IN THE TWO YEARS following the meeting at the Scottish Cultural Centre I heard Wildlife Service biologists complain about farmers whose planting schedule on land they leased from the government suited their needs rather than the needs of the snow geese. I heard farmers complain about Wildlife Service practices that seemed to hurt rather than help the birds. I heard naturalists complain about hunters, hunters complain about naturalists. I heard US biologists complain about spring hunting in Canada, Canadian enforcement officials complaining about the greed of US hunters. And, remembering that smoke-scented hall, the clustering of 'groups,' I imagined how it would be for individual men and women to find themselves in the same field on a November day when near-freezing temperatures and rain beginning to drizzle down from a dark pewter sky forced them together to share a thermos of something hot, or news of the snow geese, or, better still, something about themselves. Because it seems impossible to appreciate a subjective view unless you know the viewer, for humans are not bound to perceive as hunters, or naturalists or biologists, Canadians or Americans, so much as they are bound to perceive as individuals.

# 3

# Fall on the Fraser

IN THE COURSE of writing a novel a few years ago I visited Westham Island every month to observe the change of seasons in a rural setting: not an untouched rural setting, shades of the century before, but still relatively intact for a piece of land so close to a major North American city. I knew of the Reifel Sanctuary, at the far western edge of the island, and had visited it years before on a couple of occasions, but it was only during this novel research that my visits became regular and frequent. One day when I was driving round a curve in the road, I passed a field so white with birds the visible grass was subtle as the line of green jelly in a pillow mint. They had to be snow geese and yet to me, at that time, they were defying the nature I imagined for them by sitting so unassumingly in that old pea field, chattering like relatives at an annual reunion. This was nothing like the first time I saw them, when they printed on my mind an image full of drama, inaccessibility, wildness. In the official American Ornithology Union list, in fact, they are still called *Chen hyporborea*, the geese from beyond the north wind. How could something so wild appear to be so tame?

If these were snow geese I must have surprised them, I thought, and

I drove past slowly, the mechanical equivalent of tiptoeing, trying to make as little disturbance as possible. Laughable though it seems to me now, I remember that I hardly allowed myself to look. Imagination was about to be reined in by information.

The Fraser River rises in the Rocky Mountains near Jasper, Alberta, follows the Rocky Mountain trench west, then winds in and out of forests and drops through canyons before flowing out the valley bottomlands to finally drain into the Strait of Georgia just south of Vancouver, 1,368 kilometres from where it began. This big river creates a delta that stretches fifty kilometres across the international border into the northwestern tip of Washington State. And like the Yukon River the geese took off from in late September, it breaks into smaller channels separated by sandbars and islands before it hits salt water. One is Sea Island, site of the Vancouver International Airport. Another is Lulu Island, where the booming city of Richmond sprawls.

Small flocks of snow geese still use the outer Richmond dykes and the fringes of the airport, but airport developments, housing tracts, and shopping malls have displaced much of their habitat. By contrast, Westham Island is still rural, the marshes largely untrod, except by scientists and hunters. The relatively pristine state of Westham Island is partly the result of a provincial government policy that created agricultural land reserves and partly the result of efforts by a group of people who got together in 1960 to try to stop further loss of wetlands in the delta. The same Barry Leach who spoke at the public meeting I attended, who was at the time a college instructor and who died of cancer in 1995, and Fred Auger, a newspaper publisher who was then serving as the president of Ducks Unlimited Canada, approached the son of the original owner of the sixteen-hectare property at the western tip of the island. George C. Reifel was a well-known sportsman who made his fortune in distilling, brewing, and real estate. He used the home he built on Westham as a retreat for his family and the native and migrating birds he used to like to feed as a hobby. Since he had long dreamed of his property becoming a refuge for wildlife, his son, George H. Reifel, agreed to lease it to the BC Waterfowl Society for a dollar a year. Ten years later he turned it over to the people of Canada completely, in exchange for a promise that the sanctuary bear his father's name and that the land always be used for a waterfowl refuge. This, along with the adjoining 280 hectares of tidal marshes dedicated as a provincial waterfowl refuge, and the adjacent farm lands make up the Alaksen Migratory Bird Sanctuary, created by the Canadian Wildlife Service in 1973.

The main flocks of over 40,000 geese live out the simple rhythm of their lives here from late September or early October to January. Smaller flocks feed on the marshes nearby, north and south of here; off Sea Island and Lulu Island to the north, and Brunswick Point a little farther south, from where you can see huge ships from Korea and Japan taking on coal from the port at Roberts Bank. Then at night all the geese congregate on the foreshore marshes because tides are low enough to expose the bulrush zone, and they are safe here from prowling coyotes. They dig as deep as twenty centimetres into the soil on the foreshore to root out nutrient-rich bulrush rhizomes, and the mud has such high iron content it leaves their faces stained a rusty colour that distinguishes them from the whiter-faced California wintering geese, which forage on rice and corn growing above the ground.

A missionary who spent from 1890 to 1893 on the Alaskan side of Bering Strait, in the vicinity of Wales, Alaska, shot a goose that was, decades later, determined to be a snow goose, and probably one that wintered on the Fraser, just by the way he described it: 'A white goose with about one-third of its wings black and with curious reddish streaks about its head – looking as if some one had smeared a slightly bloody hand over it,' H.R. Thornton wrote in *Among the Eskimos of Wales, Alaska, 1890-1893*.

John Ireland, who is manager and resident naturalist of the Reifel Sanctuary, keeps a weekly log of species he has seen himself and that other people have reported to him, and posts it on the window of the sanctuary gift shop Saturday mornings. He walks around the sanctuary every day at high tide, and his quick eye spots flying, perching, hunting, diving, resting birds. In 1993, he recorded the arrival of twenty-three snow geese on the 12th of September, fifty the day after, and a further seven hundred on the 17th.

It was later in September that I made my first trip of the season that year. I had decided to pursue this meditation on observation by then, but

I had not begun any in-depth research. I was going only by what I remembered from previous years, and it was too early, I believed, for the snow geese to have come. Still relying on imagination, I remembered someone telling me he recalled that the geese always arrived on a storm. That the clouds parted like a split pillow and the geese fell as feathers out of the sky. But driving out to the sanctuary I passed a whole flock of them grazing in a potato field behind a tumbling-down, moss-covered barn where an ancient pear tree drops hundreds of little brown pears each fall. 'The expectation is great,' I noted then, 'as if the geese are bringing with them some long-awaited news from the north, but they are quite mundane, clustered on the field, rooting, chattering, sleeping, some of them, with their bills tucked into the feathers on their back.'

Fall of 1994 begins my fourth season of regular acquaintance with the snows, and I am waiting even more expectantly for their arrival, partly because I am fully into my research now, but also because the geese have become a more integral part of the season for me. So I visit the sanctuary each week in September waiting for them to arrive, waiting to see how the journey south has affected the flock.

By 13 September the harvest on Westham Island is into its denouement. I meet a truck with a load full of potatoes at the Westham Island bridge. Hand-lettered 'Corn for Sale' signs flag many roadside posts. The Little Island Farm wagon is piled with hay bales and flanked by corn stalks. There's sweet corn, dried flowers, green beans, and cucumbers for sale this week. Prices are printed on small cards, and shoppers are expected to leave their money in a tin box. Farther on, beyond a sign that reminds people to 'Respect Farmland,' green and purple cabbage make lush dusky rows against the black soil. Swallows scissor across the softly marbled grey and purple sky. The tide's in, the river high, the temperature about thirteen Celsius.

At each far-off call I raise my eyes to search the sky, but I see only Canada geese, a couple of hundred, I guess, flying from the fields out to the shore. Just before noon an island of blue spreads the white-grey rags of cloud. A high pressure ridge is holding off a storm centred north of here, the same storm that may drive the geese in. The marsh is still green, red-winged blackbirds sing exuberantly from velvety brown cattails. But the sign on the window of the Reifel Sanctuary gift shop makes no mention of snow geese for the week of the 6th to the 13th.

Next week, 22 September, it's a different story. Scanning the list on the window I spot, among others:

fifty-one greater white-fronted geese

seven mute swans

four snow geese

Four snow geese!

Summer seems to have returned. At 9 AM the temperature is already close to seventy Fahrenheit, twenty degrees Celsius. Hundreds of passerines sing, whistle, chirp, buzz in the berry bushes and alder scrub that line the path out to the marsh. A ladder is propped against the observation tower, and two men are at work on it, painting. But it's okay to go up, they assure me, as long as I watch where I'm stepping. 'Have you seen any snow geese yet?' I ask.

'We heard talk of four to six,' says the one with the cap backwards on his head. 'Somewhere out there.' He points west. I lift my binoculars and scan the slaty foreshore. 'But we heard it's going to be a bad year. The lemmings crashed all over the Arctic.'

If the lemmings crashed, then the arctic foxes that den on Wrangel Island will take more geese. So there will be fewer goslings this year. I think I see some white in the distance, but my field glasses haven't the power to bring me that close to the faraway scene.

'There's seven mute swans somewhere out there,' the painter tells me. So it might be them.

*27 September 1994.* A gorgeous sunny day ... again. Temperature nineteen Celsius. High tide at 13:00 at Vancouver harbour. I pull onto the wooden bridge that connects the island with the mainland exactly at 13:00. The river is high and splendid with a blue-brown almost satiny sheen to it. Sandy weeds and grasses sprout from mud bars: there's a wonderfully clear sky above. The truck ahead of me pulls a swaying wagon leaking wisps of hay. Ahead of it, a truck full of turnips. Out at the tower I see the ladder, mean to speak to the friendly painter again, but when I approach and ask if it's okay to climb up, the man that answers is John

Ireland, who is painting the metal grate steps from below. There's no one else in sight, the breeze is gently refreshing, the emerald marshes and blue sea stunning. It seems to me that as resident naturalist here John Ireland has one of the best jobs. He clearly relishes sharing his knowledge of birds, having watched them for forty years, since he was a child. But he reminds me that the position entails more than strolling the sylvan paths of the sanctuary, counting birds. 'I have to clean the toilets and do this sort of job, too,' he says, his conversation still rhythmic with the accent of Yorkshire, where he was born.

As we talk of the snow geese, John repeats the lemming story he heard from someone at the Canadian Wildlife Service, his neighbour across the field. 'Mind you,' he says, 'if there aren't so many it will give the marsh a chance to grow back.' I mention that they seem to be late this year, at least compared to last year, and he informs me that they usually arrive within about the same two weeks year after year.

EARLY THE NEXT WEEK I speak with Barbara Ganter, a PhD candidate here from Germany to study under the snow goose expert Dr. Fred Cooke, who is chair of wildlife ecology at Simon Fraser University. Barbara spent two weeks on Wrangel Island in July putting neck collars on geese. It's true that the lemmings crashed in most places in the Arctic this year, she says, but not on Wrangel Island. On Wrangel, as so often before, it's the weather that has had such a devastating effect. Barbara's observations of Wrangel Island, and observations the Canadian Wildlife Service biologist Sean Boyd made during his two months there in 1991, and that the Washington State wildlife biologist Mike Davison made during his five weeks there in 1993, have formed the picture I have of the place in my mind. Barbara describes a pitifully small number of goslings, all of whom appeared so small and weak she thinks it unlikely that any will make it to the wintering grounds with their families. The Russian scientist who heads up snow goose research on Wrangel, Vasily Baranyuk, said he saw yolks dropped on the grass, something he had never seen before.

I picture smaller flocks, all white, arriving here. I wonder if snow geese, who are so tied to their families, feel any sense of loss when there are no young to rear.

6 *October.* 9:30 AM Definitely deeper into fall. More leaves missing. Temperature cold enough for a warm jacket. Swags of muslin-like morning low cloud, a screen through which the light sifts.

Little Island Farm still has corn for sale, cucumbers, dried flowers, and now, huge pumpkins and turban squashes propped on the hay bales. Pumpkin patches everywhere. At the sanctuary I come across more great blue heron than usual, surprising one from its perch in a big fir at the edge of a slough. It springs off the branch with a ghastly squawk: they look more beautiful than they sound. The sign on the gift shop window says sixty-five snow geese were spotted the week of the 26th to 30th. But I can't see a single one, not in any of the fields where I usually find them. Not on the foreshore either. But the horizon is oysterish. A fog horn bellows. I can see almost nothing out there. Driving home I realize I've noticed two men with chocolate labs in the backs of their pick-ups. Hunting season opens in a couple of days and still the great flocks have not arrived.

11 *October.* Today I have company. My eight-year-old daughter Annie and her friend Miranda, on the loose during school hours because their teachers are having a professional day.

No sign of snow geese in any of the fields, but more and more elaborate pumpkin displays. At the Reifel Sanctuary the girls wade into the islands of mallards and rock doves that hang around the parking lot waiting for the seed the children like to feed them. I'm just zipping up my jacket when I hear the familiar sound, look up, and see them sailing in.

'There they are! There's the snow geese! Look!'

They continue flying over us, heading for a field I know I can see from a path in the sanctuary. Their high calls fill the cool grey air, late morning. 'Let's go!'

The urgency I feel makes no sense. They will be here more or less until spring. But I've been waiting, watching for them for a month, and they have been on my mind for years. Now they're here, and they're not mundane at all yet, but carry with them the scent of long travel: their calls broadcast news of remote places where no one has seen them, struggles no one has witnessed. And after hearing Barbara's story, I'm anxious to see if any young survived the trip south.

I hurry down the path, the girls following more slowly as they stop to distribute seed to Canada geese, a few gadwall, the beautiful wood duck whose permanent home is the sanctuary, and other birds I don't bother to notice. And there in the ploughed up potato field between Alaksen and Reifel, settled on the newly sprouting winter rye grass, are several thousand snow geese. Wonderful! We look for the grey young and see more than we expected. They did make it. They're all in a group, Annie notices, and while 'all' is a bit of an exaggeration, the juveniles do seem to be huddled mostly at the edge of the flock, close to their parents. One adult and two goslings take off and fly across to where some Canada geese are feeding. More lines come in from the foreshore. White against grey-purple sky, fluttering flags, semaphoric. We made it, they flash, we made it!

In the air they have an elegance, a drama I appreciate more now that I have not seen it for over six months. The sinuous line of their bill, throat, breast aloft, the comical appearance of their feet as they come in for the landing. This is the cartoon image, as if Scrooge McDuck is dragging his feet on the air. Very exciting. The girls go off down the path to feed the more accessible chickadees and ducks while I watch the scene and try to make sense of it. To me the young look thin, but it may be Barbara's report from Wrangel and persistent imagination manipulating my view. I see many white faces and later learn that while numerous California birds have stopped here, Fraser-Skagit birds also have white faces when they arrive. They lose the distinctive rusty stain on their face when they moult, and don't regain it until they have been feeding on the Fraser marshes for several days. This group arrived only yesterday, 10 October, Canadian Thanksgiving – and their faces are still clean. There's a satisfying feel that the seasons are progressing as expected. Houck, houck! Ou, ou! Kanguq!

It's definitely further into fall, too. The marsh, which less than a week ago still gleamed an almost startling green in the sunlight, has turned a tired brown. The cattails have split open and spilled their cottony interiors.

Later, returning from the edge of the marsh along the same path, we see that more snow geese have filled the field, and there are more people watching them; a group of Asian tourists and, farther on, at the observation spot where I stood a couple of hours ago, a pair with telescopes and note pads. It's Sean Boyd behind one, his eyeglasses pushed up onto his short dark hair, and his student assistant, Barbara Pohl, behind the other. The second Barbara in this story is also from Germany, but not yet embarked on her PhD program. It is entirely coincidental that two German Barbaras have become involved with the Wrangel snow geese.

Sean will continue field observations for about a week, training Barbara to spot and record numbers on the red plastic neck collars that the other Barbara helped to place on adult birds on Wrangel Island. This first day of the season Sean is surprised by a few geese with yellow and green collars. He has no idea where they've come from until he learns later this week that they belong to a small colony nesting on the Russian mainland, which Japanese biologists have collared.

He estimates that there are 10,000 to 15,000 snow geese in the flock we're watching, and maybe 1 to 5 per cent are juveniles. There are lots of California wintering geese, mixed in at this point, too. So his census of Fraser-Skagit birds will have to wait until the entire flock has landed and sorted itself out. As well as his many duties as a CWS biologist, Sean has been trying to finish his dissertation on the winter ecology of snow geese. Their arrival has taken him away from his desk, he complains, half-heartedly it seems to me. 'But it's nice to have them back.'

John Ireland has other things to do, too, such as finishing the painting of the observation tower. The weather is clearly changing, this may be one of the last dry days for a week. But farther along the path, at another observation post, he too has his scope aimed at the field and is copying down neck collar numbers. 'There were 500 or so in on Sunday,' he reports. 'Then late Monday, Canadian Thanksgiving, the big flock came in.' I don't know if it's my maverick imagination at work again or if I am correctly observing the fact that Sean Boyd and John Ireland, who have been at this for years, feel as exhilarated as I do. The snow geese are back!

And they are already settling in, grubbing on the marsh, beginning a winter of socializing, finding partners, and grubbing for the rhizomes or tubers of *Scirpus americanus* – three square bulrush – whose profusion here, almost at its northern limit, is one of the reasons why they congregate

Snow geese coming in for a landing on Westham Island

Snow geese families grazing at the Reifel Sanctuary

on the Fraser. 'It's kind of a neat plant,' Sean Boyd explains, 'because it grows in soft mud and has big roots, high in nutrient value.' So it is both easy for the geese to get at and worth the trouble.

Out on the foreshore they are also safe from terrestrial predators such as the coyotes that have taken up residence on the island in recent years. Bald eagles can and do take the occasional goose, but they cause more trouble simply by disturbing the flock, swooping over in a search for injured birds, causing the flock to rise up excitedly and lose some of the calories they spend all day ingesting. Interestingly, when a great blue heron, with its impressive six-foot wing span, flaps over them, they pay little attention.

The geese feed in farm fields, too, more often in recent years than they used to, according to farmers who grew up on Westham Island and claim that, historically, the snow geese came into the fields only in years when they had many young. Since Sean Boyd has found that the marsh still produces 80 per cent of the biomass it would produce if snow geese were not feeding on it, why they feed in the fields when there are sufficient rhizomes in the marshes is still a mystery. Some scientists have speculated that field feeding might provide more value for money: it might be easier to forage in fields. Or, it might indeed have something to do with the young. Barbara Pohl, who by the end of the season will have spent days and days watching geese and recording neck collars, has noticed that it is definitely those groups with young that return again and again to the fields, even after being frightened out of them by hunters or other predators. In the fields they congregate in family groups, grazing on forage grasses and leftover potatoes. When they are not feeding they sleep or preen, using their bill to squeeze oil out of the gland at the base of their tail and distribute it over their feathers. Preening ensures that their feathers stay waterproof and also keeps their bills in good condition. A number of erect-necked, sharp-eyed geese, generally the ganders, keep watch while others in their group tuck their bills into the base of their wing and rest. If most of the group is sleeping, it is easy to mistake a distant field of them for stones. And on days when the land is sodden with puddles and ponds that reflect the sky like so many mirrors, it is even harder to tell if the geese are geese, or only the illusion of geese.

Snow geese drinking

When close enough to be sure they are geese, even the seasoned wildlife watcher has a hard time guessing how many individuals make up the flock. Sean Boyd says that most people usually underestimate by half. 'If you think you see 3,000 birds, it's probably 6,000.' And counting, or at least estimating, is instinctual, it seems, when confronting numbers like this. It's the first easiest way to make sense of the sight.

Because we know that snow geese travel in families, it is logical to assume that a white pair with grey young are a family. But if both white geese are resting it is almost impossible to tell which one is the female, which the male, though when they are standing the male clearly stands taller. Determining which sub-population they belong to can be even a bigger challenge, especially early in the season when the faces of all the geese are white. Russian biologists believe that Fraser-Skagit male snow geese are larger and their bills are longer than California geese, which, because they feed above ground, don't need long bills. But when birds are being banded on Wrangel Island, biologists decide which are Fraser-Skagit geese and which are California geese on the basis of facial stain. They have a light-to-dark scale of one to six, the white or lightly stained birds being those that winter in California. 'The threes and twos, we're not sure where they're from. But the longer they stay the more sure we are that they're Fraser birds,' says Sean Boyd.

*18 October 1994.* Fog thick around the river. About ten degrees Celsius. The tide is not high but at Alaksen I do see some snow geese, a small number, twenty-three, with ten young, feeding and resting with Canada geese. A man in a white car, with the licence plate MYNAH, and a scope mounted on his window, is parked alongside the field.

I don't see the main group of snow geese until I am almost round the marsh path. There, to the south, a great flock of them bleach a section of the foreshore. Their calls wash back across the thick grey air to where I stand on the observation platform, which has a wonderfully isolated feel to it in the fog.

Driving back out I see the same white car is parked alongside the field where the same few snow geese are feeding and resting. I walk up to the driver's window, thinking I'll ask to look through his scope, and am quickly caught up in conversation with seventy-six-year-old Roy Phillips, whose dashboard is littered with lists of collar numbers. He's been recording numbers for seven years, he tells me, and drives down to the Skagit every week to record the numbers of collars he spots on snow geese there. A retired employee of the Vancouver Parks Board, he travels out to Westham Island almost every day this time of year. 'Oh you have to have time,' he says. He also says they come closer if he talks to them like an Eskimo. How's that, I ask, and he demonstrates, hand cupped around mouth, voice high and throaty. He thinks this small group has been led here by the juveniles. 'The young want to move, and so the parents have to follow.'

This seems to add weight to observations local farmers and the student biologist Barbara Pohl have made. More experienced scientists confirm that adults with young do tend to stick together, either in isolated groups, or on the fringe of groups, where the grass in the field is likely to be in better condition.

10 **November 1994.** Fog again on Westham, some of it hovering in an eerie layer just above the ground. As I drive out to the sanctuary, I see big flocks rising from the edge of the marsh where hunters' shacks make Monopoly-house silhouettes. I hear a shot. The geese fly across to Alaksen, then fly back in u's and v's. The lines change shape as the geese cross the fields, disappear into vapourish sky. In the entrance field, I notice a single goose, an adult, far from any Canada geese. All by itself. Why? Could it be injured?

As fall progresses there are more cars in the parking lot, more people strolling the paths. Many seniors, but also mothers with children, preschoolers. As the geese home into the refuge I overhear one elderly woman remark to her companion: 'Imagine how they know that this is a safe place for them, a sanctuary.'

Midway down the path to the marsh I stop to watch as an enormous throng flies directly over me, quite low, as if they might brush the tops of trees. They look healthy, beautiful. The shape of the line changes as each bird flaps into a new position. Such a thrill! For many minutes the whole blue sky overhead is wing-to-wing geese. The odd one, even a pair, tries to cross the flock and what a sight that is ... to see any one of these trying to act as an individual against the force of the crowd.

They settle in what I have come to call Thanksgiving field. Landing, their wings come up, shoulder muscles strain, tail makes a fan. They dance down like an airplane on tarmac. It's cold today, perhaps seven Celsius. Hurricane-force winds on Tuesday ripped most of the leaves from deciduous trees.

On the far edge of the flock I see one goose stretching its neck towards another. Find your own food, he might be saying.

A man, practising with a video camera, points out how the flock thins at the edges, to make room for more geese, he theorizes. He tells me that he always visits the refuge on 10 November because it's the day he first saw a bald eagle.

Another group of perhaps a couple of thousand flaps in, calling. The assembly on the ground calls back. And they call again when a black-crowned night heron flies low over them and lands at the edge of a slough. Terry Vandersar, a columnist for the *Vancouver Sun*, has been watching snow geese for years, and this season he began filming them. He believes he can make sense of some of their calls now, and related a story of one day's experience: 'Several thousand snow geese had resettled in a field of grass after being startled into flight by a bald eagle. One grey juvenile, flecked with white feathers, had become separated from its family. This bird remained in the air and circled the flock calling loudly in its adult voice. On its fourth or fifth overflight, this juvenile passed by my camera quite closely. As it happened, a family of geese on the ground nearby answered vigorously. This family group of two adults and two juveniles responded a second time when the juvenile passed overhead again.

'The crowded field forced the juvenile to land some ten metres away. It ran a gauntlet of angry adults while calling for its family. To make a long story short, the five geese were reunited on the ground. Almost as soon as the lost child found its parents and siblings, the juvenile put its head down to graze and started its soft litany of: "I'm okay; doing just fine."'

# 4

# *The Scientific View*

ONE MORNING in the summer of 1993 I heard on the radio that the renowned snow goose expert, Dr. Fred Cooke, had just been appointed chair of an innovative program in wildlife ecology research, cooperatively supported by the Canadian Wildlife Service and Simon Fraser University, which sits at the top of the hill mere miles from where I was living at the time. A snow goose expert in the neighbourhood? Terrific! Who better to provide the scientific view than a man who has made snow geese his life's work, who has the biggest set of data from one of the longest-term studies of birds in the wild in North America, maybe anywhere. He must know everything, I thought.

The plot thickened in an even more interesting way when I called SFU and learned that there was a Dr. Cooch on the team of biologists that had accompanied Dr. Cooke west from Queen's University in Kingston, Ontario. Cooch? Could it be? The very same Cooch whose name I saw cited in virtually every piece of literature on snow geese I had come across in the library? Had I, by the sheer accident of my early morning radio habit, struck research gold?

The answer is more or less yes. Not only does Dr. Cooke know most everything, the Dr. Cooch who works with him, Evan Cooch, is the *son* of the man whose name I saw so often in print during the library stage of my research. F.G. Cooch – Graham – former explorer, arctic ornithologist, then senior scientist and senior adviser to the Canadian Wildlife Service, now a professor at New Mexico State University at Las Cruces. Even more serendipitous, it turned out to be Graham Cooch, during a talk he delivered in Kingston in 1967, who inspired Fred Cooke to begin studying snow geese in the first place.

It seemed I had stumbled upon a kind of family tree of snow goose specialists who were not only accessible but also agreeable enough to help me fill in this part of the picture. And Fred Cooke kindly offered the names of others on his team who might be of help, most notably Barbara Ganter, who interested me because of her gender. For in the research I had conducted thus far, I had talked almost solely to men, and it worried me that the views I was accumulating would be skewed in this respect. The only drawback of these four – Cooke, Ganter, and Cooch, father and son – was that until 1993 they had worked with snow geese that nest in the eastern Canadian Arctic and migrate south through the central flyway to wintering grounds in Louisiana and Texas. I wanted to concentrate on the geese that nest on Wrangel because these were the birds I personally saw each fall.

Though as time goes by Fred Cooke's team has had and will continue to have more and more involvement with the Wrangel geese, there's been one person at the Canadian Wildlife Service whose preoccupation stretches back to 1987, the habitat biologist Sean Boyd, whose curiosity about snow geese grew out of his studies of the marshes on Westham Island.

So here they were, the individuals I cast as scientific observers. Instead of a single view, I realized quickly, it was fated to be many faceted, the view of snow geese as a compound-eyed bee might perceive them.

After the hours I spent with them in their offices, homes, in the field, I came to think of them as species of the genus scientist: the silver-crested population geneticist being Fred Cooke; the spectacled number cruncher, Evan Cooch; the long-necked behaviour ecologist, Barbara

Ganter; and the elusive, many-hatted habitat biologist, Sean Boyd. Still another variation on the theme turned out to be Mike Davison, who trained as a wildlife biologist and is now responsible for the management of snow geese as well as all the other wildlife that inhabit the Skagit Valley. Graham Cooch is the granddaddy of them all, and after sixty some years studying birds and managing various populations, snow geese are still his favourite birds.

These six are a small proportion of the total number of biologists in North America who study arctic geese: nearly two hundred attended the arctic goose conference in Las Cruces, New Mexico, in January 1995. So why all the interest in geese, I wanted to know.

According to Evan Cooch, snow geese are theoretically useful species and lend themselves to all sorts of interesting work because they are high density nesters. 'So you can generate lots of numbers very quickly.'

Then there's the fact that, within the Wrangel population, birds that winter in California and birds that winter on the Fraser and on the Skagit sometimes interbreed, presenting complex management challenges.

Snow geese also make an ideal study species because they are more or less accessible at all times in their life cycle, so they can be easily counted; and they behave as populations rather than as individuals. On top of that, they are grazers. 'You can look at food intake rates and varying food supply much more easily with a grazer because you can measure how much they eat. Whereas with an insect-eating bird you can't find out what they're taking,' Fred Cooke explained.

'We're really interested in the population dynamics of all bird populations; it's just that we can do it well with the geese. They are valuable in their own right, but they are valuable too as indicators of things going wrong with the environment.' For example, it is easier to see the effects of pesticides on birds than on insects. The higher up the food chain, the more concentrated the damage.

We're talking in Fred's Westham Island office, Pacific headquarters of the Canadian Wildlife Service, now called the Pacific Wildlife Research Centre. The change in name reflects a difference in how the people responsible regard wild creatures. What began as proprietary, managing animals to maximize their usefulness to humans, has mutated into an interest in animals for their own sake. The changing study emphasis reflects a change in values and a change also in how science is done.

'There was a time when looking at animals in the field was what most people did. Then there was a time when looking *inside* animals and writing them up and looking at their chemistry became somehow much more rigorous. This was when people understood the chemistry of DNA. There was a strong sense in biology that the right sort of science was to try and explain natural phenomena in terms of the chemistry and physics of it, that that was somehow harder science. That sort of thinking still exists, but it's not the sort of science that ecologists do; it's what molecular biologists do. When I went through university there was a sense that somehow ecology as a discipline was much more loose and sloppy and soft than the real science which tried to explain things in chemical and physical terms. That's something I've always resisted but it certainly was popular when I was in graduate school.'

The bleep of the phone interrupts him, and he excuses himself to take a call, the first of several that punctuate the morning. One of the projects he is overseeing as chair of wildlife ecology is the establishment of a permanent research station on Triangle Island, off the northwestern tip of Vancouver Island, and the arrangements he is having to make – funding, transportation, facilities – expose the nonscientific part of this scientist's work. After each call, however, he hesitates for only a beat before gathering his thoughts and continuing.

'There are two fundamental frameworks for biology. One is the one I described already which says, let's try to understand these things as much as possible in terms of chemical-physical processes. We can understand how a bird feeds by really understanding intake rates, and digestion, and by understanding the actual chemical processes that make up the body. And there are others who say let's try to understand their role in terms of what we know about evolutionary processes. Given that Darwin was basically right, what predictions can we make about the way in which an animal behaves, how it avoids predation, things like that.

'In both frameworks you can formulate hypotheses, and you can do good science. Mine happens to be the evolutionary framework. Most people who are into evolutionary questions put themselves into a

Fred Cooke

framework of either genetics or ecology. There are population geneticists who look at the way genes affect working populations, why different populations have different genetic characteristics. And then there are population ecologists, who are interested in demographic processes. Ecology itself is often split up into what used to be called synecology, where what you're interested in is a whole ecosystem and how bits of it work together. And the other used to be called autecology where what you're doing is concentrating on a specific species and how it responds to all the environmental circumstances that it faces, including the other animals. So it's not completely dichotomous, but you tend to be one or the other.

'People who are into habitat are usually more interested in how all the little bits fit together. People who are into population ecology are usually more interested in how one or a few species work together.

'Then there are behaviour ecologists that ask how the bird behaves in response to the real world, behaviour geneticists, ecological geneticists. People like to classify themselves in various specialties. But I think of it basically as a four-pronged approach: you're interested in the ecology, you're interested in the genetics, you're interested in the behaviour, you're interested in the physiology. Physiology is how these birds work. And you're probably interested even in morphology, what they look like, although that's an area that's investigated, people know about it now.'

A knock at the door halts the flow again. He apologizes, explaining that two separate people have booked his time this morning and he has ended up with schedule conflicts. He manages to delay one meeting until after lunch and turns his attention back to the subject at hand.

'I don't know what I'd classify myself as any more. I've done behaviourial work, ecological work, of course genetics. I look at the interface between the three. There are a lot of good synthesizers, but increasingly a synthesizer is thought to be so much of a generalist that he doesn't have the specialized knowledge required for any particular questions. So I would say that people who are good synthesizers tend to be less highly regarded in the university environment than they should be. I had a colleague who was extremely good in talking about any subject. He was not always totally accurate, but he saw the big picture. Because of that he was never regarded as "the expert on." Therefore he didn't have the same scientific reputation as some of my other colleagues, which is sad because he was good.'

Evan Cooch tells the following story about Fred Cooke. 'Fred and Robert Rockwell are the two major players in the La Pérouse Bay project. And they just finished the big book, the final result of twenty-five years' worth of data collection and analysis. For a long time we had a standing joke as to what the title of the book was going to be. We decided that Fred's title was going to be one of two things, *My Friend the Snow Goose,* or *Never Cry Goose.* And that represented at some very sarcastic level our view of how Fred sees things, because he really is a birder. He just loves it, and he's good at it. Whereas Rockwell is a complete reductionist. His title was going to be *Population Biology of the Arbitrary Study Organism X.'*

The book by the way, which was released in the spring of 1995, is called *The Snow Geese of La Pérouse Bay: A Study of Natural Selection in the Wild,* and its third author is another associate of Fred's, David Lank. But I had talked to Fred several times before I met Evan, and it was easy to appreciate the story. A white-haired, white-bearded man of modest stature, whose gentle tones bespeak a County Durham, England, beginning modified by Cambridge and then central Canada, he and I first met on a day when he couldn't keep the appointment we'd made at his office because he was having trouble with his back. We sat in the garden of the home he and his wife Sylvia had recently bought in White Rock, a few minutes north of the Canada-US border, and as we talked he paused to point out various birds that appeared, including three red-tailed hawks circling on the columns of air that tower up from the coastline.

Evan Cooch was telling me that when Fred Cooke looks at birds he sees birds, whereas when Rocky Rockwell looks at birds he sees numbers. So what makes for the difference between a Fred Cooke and a Rocky Rockwell? Did background explain it? Was there something in the genes of this geneticist that determined he would become a birder? 'I suppose I was interested in birds from the age of five. In Britain, maybe more so than here, if you're a kid you're supposed to have a hobby. My hobby was bird watching. So I joined the local natural history club, and at school there was a good naturalist's society.' His parents enjoyed cycling in the countryside around Darlington, in County Durham, where Fred was born in 1936, and knew the common birds. But his father was a plumber, and his interest was incidental, while even as a child Fred had the urge to go further, to look for nests and find out the names of birds that were not so common.

'You'd see things, oh there's a chaffinch and then you'd see one you didn't know, so you'd buy a bird book and then you'd suddenly see pictures of all the birds you didn't know so you'd try to find those. I remember getting to know things like, well finding a grey plover. That was a rare bird to me, because I'd seen it in bird books but I'd never seen one in the wild, so I was walking along the seashore one day and all of a sudden there's a grey plover.

'Then naming isn't satisfactory. You want to know why they're there, you want to know what they're feeding on, who's eating them. When they start to nest. Then you start asking more questions about the birds, then you start reading. Monographs. I remember very early in life reading some of these books that were an attempt to find out as much as you could about a particular species. I remember a fellow wrote a book on the redstart, a bird I'd seen occasionally, but he'd done far more. He'd actually been in a German prisoner of war camp, and he'd made some nest boxes and stuck them around, and he got these birds nesting in them and he therefore knew all about how many eggs they laid and how successful they were and what they fed on.'

He coughs and pauses to clear his throat, then continues, recalling that he must have been about thirteen when he started taking notes on observations he made with his natural history group at the Quaker public school he attended, Bootham School. 'I used to write a diary, and I had photographs and I would note changes in duck populations over the winter. Then we started bird banding at school. There was a very good biology teacher. We'd go out early morning and make traps for starlings, band them, note the sex of them, whether they were young or old. You hope that someone finds them somewhere else so you know where they've gone. So you collect information that leads you to a sense of wanting to ask questions. This biology teacher got us catching starlings, and then he got us into finding starling nests, and we did a study of our village, where we plotted all the starling nests and tried to figure out where they spent the night. So we all went to different parts of the town and at night when they were heading towards the woods we drew lines and from that we figured out they were using the woods at the south of the town, and we went and found these hundreds of thousands of starlings and started catching them and banding those. That was when I was about seventeen. I spent most of my leisure time doing that sort of thing.'

Cultural traditions, of course, inform perception to a major extent, and the tradition of bird watching in England directly affected the views of Fred, as well as others who enter this snow goose story and, in fact, the practice of science itself in Canada.

As Evan Cooch pointed out, 'There are really two ways of working in the field, depending on how you do science. In the scientific community there's the question of long-term, short-term studies, but even within either of those there's the question of manipulative science versus what's called mensurative science. Observational. Manipulative science believes that the way to test certain hypotheses is to do the equivalent of a physics experiment where you actually manipulate the system you're looking at. Changing the number of eggs that are in a bird's nest, for example. Or, if you think that the colour of feathers makes a difference, actually painting the bird. Actually physically changing the system from normal. Now that's one approach, and it's a fairly common approach, especially in the States. But we don't do that, what we do is the latter of the two, the mensurative approach. Observational, which is what I call typical of the British tradition of the sort of one or two individual people who like this particular species of bird that nests in their area and they just watch it year after year as a hobby for fifteen million years. But they don't manipulate, just record religiously everything about it. And to a large degree, that's what the La Pérouse Bay project involved. An army of people going up to the colony every year, collecting lots of pieces of basic information year after year.'

It's noteworthy that one of the two men who trained Fred in how to collect information about snow geese was the British ornithologist Ian Newton. The other was a Canadian, John Ryder. Both men were bird experts while Fred, when he first got involved with geese, in 1968, was working as a geneticist. For, as well as being interested in birds, he was interested in plants and actually wrote his dissertation on the genetics of fungi.

'It was a diversion. I thought I should take a job where there seemed to be a lot of usefulness. Two areas seemed important: plant diseases and plant breeding. I thought birds were more of a hobby than a career. More of a pleasure. Nobody would ever pay me for studying birds.' It was not until several years after he came to Canada in 1963 and accepted an appointment in the biology department of Queen's University that Fred Cooke found an opportunity to work his lifelong love of birds into his career.

'That was quite an event in my life. I was doing my research on the fungi and not particularly enjoying it. And then someone came along called Graham Cooch who gave a talk on snow geese at the local natural history club, and I'd been asked in advance to thank him on behalf of the club after his talk. At one stage in his talk he said there's something interesting about the genetics of this bird. My data don't make sense on that point but I know they interbreed. Afterwards it struck me that it was an interesting slight extension of what I was doing. I was breeding fungi together of different colours to see what the offspring looked like. I said to him, I think it does make sense if you make additional hypotheses, and so he said I'll send you my data bank. I looked at his data bank and sat and thought about it for a while and talked to various people and came up with a little twist that would explain his data if that twist was correct when it was tested. The twist was that the birds didn't choose their mates randomly but chose them on the basis of who had brought them up. The idea was that if your parents were dark-coloured you mated with a dark-coloured bird. If your parents were either colour you didn't care, if your parents were all white you chose a white-coloured bird. Now once you add that to the genetic model it began to fit a pattern. So that we could test it, go out to the field, throw some bands on the chicks, and see how they came back. See if the white chicks came back with a white mate, so we did that and it worked out.' He says this modestly, understating the significance of his findings. For what he and his team discovered was that the behaviour of the birds was influencing the genetic structure of the population. 'That was a new generality at the time. Something biologists had not thought. Genes affect behaviour, but the thought that behaviour affects genes was a new idea.'

Here I should point out a big difference between the geese that nest on Wrangel and the geese Fred Cooke and Graham Cooch studied in the eastern Canadian Arctic. With rare, extremely rare exceptions, the Wrangel Island population of snow geese consists only of white-phase birds; in the eastern Arctic, snow geese come in two colours, white with the distinctive black-tipped wings, and 'blue' snow geese, which were long thought to be a different species. In Graham Cooch's time he and other biologists discovered that the blues were not a distinct species, but rather, they thought, a subspecies. 'It was not until Fred Cooke's extensive genetic work that we learned the blue and white are merely different colour phases of the same species. The interbreeding between the whites and blues, the gene exchange between them, determines that they have more similarities than differences,' Graham Cooch explained.

But all this and much more was to be found out later. Whether he knew it or not, Graham Cooch was handing a kind of torch to Fred that day in Kingston, one that he had taken himself as a young student at Cornell University from his graduate adviser Oliver Hewett. 'Hewett decided that since I was Canadian, and I had a lot of field experience, and because the amount of information available on the blue phase of the blue-snow complex was really very incomplete, that I should go to Southampton Island and work on my doctorate.'

In his late sixties now, with arthritis and other health problems making life difficult for him, Graham can hardly believe himself the lengths to which he went to learn about snow geese. Southampton Island lies just below the Arctic Circle at the north end of Hudson Bay.

'If you wanted to be a goose biologist in my era you had to be prepared to live under very primitive conditions. This never bothered me or anyone else who worked in that part of the world. It went along with the turf. Looking back I think we were in a place like Boas River with no maps, no photographs, and no contact with the outside world for four months at a crack ... It scares me.

'At that time the only way to get to the breeding grounds was to come in the autumn before, spend the winter, and then proceed to the colonies by dog team in the summer, get back to civilization and get the boat out in the fall. I was able to fly in a Canadian Air Force transport squadron from Churchill to Coral Harbour, and then with the Eskimo that I worked with and his family, Harry Gibbons, better known as Umalik, I went by dog team from Coral Harbour to Bear Cove then overland to Boas River. First it was my intent to camp at the site utilized by Bray. The Eskimos said that it would flood, and we moved westward to an area that nearly flooded but didn't quite, about eight miles from the delta proper of the Boas River.' He was searching for a large colony of snow geese whose existence had been known for a hundred years.

'In 1952 when I made my first trip into the country, and it took five days to get there by dog team, the map showed unexplored territory, a few dotted lines. We had no aerial photographs: there were no details of the area. I did not know the precise location of my campsite until about 1986 when I landed by helicopter. It wasn't where I thought it was, but much closer to the coast. It was on an island. And the sites I had used to capture geese for banding showed up and then I saw a dam I had built to provide fresh water, and we landed. But I did not know for over thirty years exactly where I had been located in the Boas Delta area.'

Interestingly, although they ended up concentrating on the same species, both committed to science, Graham Cooch's involvement with birds began in a much different way than Fred Cooke's. He first saw birds as quarry.

'My earliest recollection is my grandfather shooting, he was crippled, and shooting sharp-tailed grouse off stooks in Saskatchewan wheat fields near Battleford. My job was to go out and get the dead birds.'

But his family soon moved to Ottawa, where, although he was to come into contact with some of the most important names in the history of ornithology in Canada, he claims that 'the first really formal piquing of my interest occurred in Hopewell Avenue Public School in Ottawa, with Miss Myrtle Melbourne. She had lists posted at the back of her class of various plants as the children brought them in and first sightings of birds. And because of my competitive nature I concentrated on the birds.'

Walks along the Rideau Canal took him past the house of Percy Algernon Taverner, ornithologist for the National Museum and author of the first *Birds of Canada*. He joined the Ottawa Field Naturalists and began visiting the National Museum every Saturday, learning from Dr. Austin Loomis Rand how to measure and mount birds. When Dr. Rand left for Chicago, to take the post of curator of ornithology at the Field Museum of Natural History, Graham continued under W.E. Godfrey, author of the second *Birds of Canada*.

'The two other influences at that time were with the Canadian Wildlife Service, Harrison F. Lewis, who became chief, but who was at that time the migratory bird officer for Ontario and Quebec, and the previous director or chief, from 1918 until his retirement in 1942 or '43, Hoyes Lloyd. They also used to encourage me, and I used to meet Lewis in particular along the Rideau Canal on Sunday mornings when he was out bird watching. I had previously met his wife in the Ottawa south library. I was in borrowing some bird books and she said are you interested in birds, and I said yes and she said then you should come and talk to my husband. I was so presumptuous I phoned him and he spent a couple of hours with me one Saturday morning going over my notebook. He was a very exacting individual and would point out the errors of my ways and what I should learn.'

As he reflects on his career, more and more familiar-sounding names roll off his tongue, names I've seen cited in literature on ornithology, wildlife science. Names that now appear on maps as the names of rivers and islands. Dewey Soper, Tom Manning, Charlie Gillam, John Lynch, to name just a few. During the McCarthy era in the United States he was the principal liaison between the US Fish and Wildlife Service and the Soviet Union. From the early 1950s he worked with US Fish and Wildlife biologists and representatives from various state agencies on the Pacific, central, and eastern flyways, and rose through the Canadian Wildlife Service to become the man who ultimately decided, on the basis of population studies, what species could be hunted in Canada and when.

Although European-trained wildlife biologists, from countries where wilderness is virtually non-existent, have long valued wild creatures simply because they exist, the tradition in North America has been different. Pioneer European settlers followed the example of the Native people they encountered when they arrived in North America and lived off the land, sustaining themselves with the fish and birds and mammals that seemed so, and were then, abundant. The hunting tradition exists still, most strongly in rural areas. And the purpose of a good deal of the population survey work that wildlife biologists do is to determine how many of any given species may be hunted. But because of his involvement with scientists such as Taverner, Godfrey, Lewis, and Rand, and because of his training at Cornell, Graham Cooch always recognized the intrinsic value of science.

Fred Cooke believes that projects such as his long-term study at La Pérouse Bay would never have been funded had it not been for Graham Cooch and Hugh Boyd 'who were very anxious to see good basic science carried out on goose populations and helped get support for us. We'd often talk over the ideas and Graham always kept an active interest in them.'

'For about twenty-five years I was an Ottawa bureaucrat,' Graham says, as he approaches the end of a long session on tape, 'but I never gave up my interest in science. The thing I'm proudest of was the creation of the CWS scholarship program and at the moment, nearly every faculty member who teaches wildlife management or ornithology in a Canadian university, or ecology in a Canadian university, was a beneficiary of that program.'

That would include his son Evan, who took a circuitous route to end up in his present position as a member of the faculty of Simon Fraser University. Though he had the opportunity to learn from a master field biologist, Evan didn't think of trips into the bush with his father as anything more than an excuse to do something fun outside, to come up with a story to tell at school about something he'd done that nobody else had done.

'Other than that I didn't have much interest. My mother is also in science – she was in physiology – and I think at some levels I leaned initially towards what she did because it was more obvious to me as a kid. She taught in medical school.'

Because Graham Cooch lives in New Mexico and I'm in Vancouver,

I've been communicating with him by phone and tape, imagining the stories he is trying to relay in verbal shorthand, memories of months and years spent in places on the map of North America that are virtually empty of print. My talk with Evan takes place in his office at the SFU campus, a sprawling compound of cement and glass perched atop Burnaby Mountain, where the choicest spots offer views of Burrard Inlet and the stunning blue range of Coast Mountains that rise just behind the skyscrapers of downtown Vancouver. None of this is visible from the South Sciences building. The shades are drawn in any case, and Evan, who is thirty-four, sits in front of a computer screen. It's summer and he's dressed in shorts, sandals, and a sweatshirt from New Mexico State. When he tells me later that he likes to work out for recreation, I see the reason behind the muscular upper-body build of a weight lifter or swimmer. But he also tells me that he likes to read literature, and evidence of that comes through in the interview too as he reflects upon the transitions between himself and his father, the linkages between himself and the other scientists he works with. His literary bent is confirmed by Barbara Ganter, who tells me that Evan won a short story contest at Queen's.

'My dad is sort of the romantic, old-school arctic explorer naturalist, compatriot of Aldo Leopold. He was a bird watcher from the womb. When I came into this project I might have been able to recognize five or six different species of birds. I was not even close to a birder. There was a separation with what you did at work and what you did at home. Thing is, with my mother, things would happen and we'd talk about it. If you got sick, you got a medical explanation. You say your dad works with birds and people get this vision of this guy in shorts, with a pair of binoculars and a funny hat.

'I didn't even take any ecology courses, I took physiology courses. It was really what I wanted to do. I was good at it, but I seemed to have a lot more fun and a lot more ability at recognizing patterns, just as a general skill, and I found more and more that I liked playing with numbers, not as mathematics, but as statistics, applied statistics. And I liked in this mass of numbers finding relationships. Then I realized that what I was talking about was population dynamics. Fred had the biggest set of data in the world and I knew, just from growing up with my dad, basic goose biology, so I came in, working on the exact same species as my father. But the questions have changed, and the way of doing things has changed. My father always accuses me of reducing this magnificent species to a bunch of numbers. It's not entirely fair, but it's probably more accurate than it should be. And Fred had a different perspective, so the whole development of the way we look at geese as a study organism, and just as something interesting to look at, has changed.

'I'm somewhere in the middle. Somewhere closer to Rockwell than to Fred. Fred would say I'm not even remotely close to him. And I think that's the way this project has worked, there's the balance of various sciences.'

Long before I met Evan Cooch or Fred Cooke, a couple of years before, when I was still working on the novel that took me out to Westham Island on a regular basis, I met the Canadian Wildlife Service biologist Sean Boyd. It was 1990 and I had stopped into the CWS office to ask an acquaintance, the seabird specialist Gary Kaiser, some questions about the local flora and fauna. When I inquired about snow geese he said the man I wanted to speak to was Sean Boyd. I found him in a room not unlike the one where I interviewed Evan Cooch, except that Sean could step from this one out into a field, lift his binoculars, and see the snow geese that were represented on his computer screen as collar numbers. I remember him being surrounded by paper, printouts. I remember feeling I was intruding on the time of a man who was totally consumed with keeping up.

Sean was interested in looking at the impact of snow geese feeding on the bulrush that, with cattail, forms the green collar around Westham Island. The question he wanted to answer was this: how does the snow goose feeding affect plant growth, and as the rhizome, root, mass of the bulrush is knocked down, how does it affect the food supply of snow geese?

Born in 1950 in rural Nova Scotia into a family that had no particular interest in wildlife, Sean earned his BSc at Dalhousie University in Halifax. 'I actually went into engineering for three years and decided I didn't want to be an engineer for the rest of my life, so I tried biology.' He completed his masters in biology at the University of British Columbia and shortly thereafter began the studies of the marsh that got him hooked on snow geese, work that became so consuming he chose it as the subject of his PhD dissertation, which he was frantically trying to complete in the summer and fall of 1994, while simultaneously directing continuing fieldwork on the snow geese and other birds, and trying to leave some time for his young family. Sean spent two months in the summer of 1991 on Wrangel Island.

In 1994, Barbara Ganter, who was then twenty-nine, spent two weeks there. While it would be stretching it to say that Barbara resembles a goose, she is tall and long necked, and she jokingly admits that she chose to study geese because her last name is Ganter. Gander, in German. Barbara comes from Bonn, Germany, and was engaged in studies of barnacle geese when she heard of Fred Cooke and wrote to ask if she might join his program at Queen's University.

The only child of a doctor and teacher, she was not particularly interested in birds growing up, nor is she now, not in the sense Fred Cooke and Graham Cooch have been. But like Fred, and Sean Boyd, she originally concentrated her studies on plants.

'I was all set to do my thesis on botany and I said to myself, you've done all this and you've never really been outside. So I ran off and did half a year of fieldwork with the World Wildlife Fund in northern Germany. At the end of that time I decided I didn't want to do plants any more, I wanted to do birds. There were quite a few projects to choose from and one of them was barnacle geese and I liked it best.

'I also like snow geese a lot. Everybody likes their study species, but I'm not a birder. I like looking at things, but I'd never run around and try to find as many species as possible.

'Fred would not go around the world to find rare species, but he's definitely a birder. Just this afternoon when we were about to go home we were standing there looking at the snow geese, and Fred came by and all of a sudden he got extremely excited and he said: "What's that bird sitting there on the wire?" and he grabbed my scope, I was looking at the geese through my scope, and he grabbed my scope to look at that bird on the wire because it was some weird kingbird. He got extremely excited about this bird. And I would have thought okay it's a bird, I don't really know what it is but I don't really have to know either. So it's a difference in attitude. He does get excited about rarities whereas I don't. The other day at the Reifel Sanctuary somebody said there's a sharp-tailed sandpiper out here, they're really rare, and so I said okay, there's a sharp-tailed sandpiper, so what?'

In high school Barbara liked languages and had planned on an academic career in linguistics, but her Grade 12 biology teacher instigated a sharp turn in direction.

'She was still in training so she was very young and very enthusiastic and we did some exciting stuff while she was teaching us. She took me to the university in Bonn one day and we sat in on a course and they were identifying bird skins, which I thought was extremely exciting.'

In reply to a question I have about the difference between how men and women scientists work, if there is a difference, Barbara answers cautiously. 'There seems to be a tendency that if you look at what women have done as opposed to men, women are more interested in behaviour and looking at geese than in getting 20,000 egg measurements and putting them through a computer. But it's just a tendency and I don't know how much it applies.'

There's much more of a difference, she says, in the way Europeans and North Americans view geese, because in Europe 'there's not much nature left and they're a lot more careful. It would be unthinkable to do a study where you had to shoot 200 geese and grind them up to get the data. It's interesting, because I use the data, but I'd never want to do it and in Europe you wouldn't be allowed to do it. So that's one difference, the studies in Europe are nonconsumptive studies, whereas there are consumptive studies in North America because there's a lot of waterfowl management going on over here. There's a lot less in Europe because there's none to manage. Waterfowl populations are protected there.'

CONSIDERING WHAT IS KNOWN about the subjective nature of perception, it seemed unwise to accept reports on snow goose observations without knowing the observers. This is why I considered it important to introduce the people who provided me with the observations that make up such a large part of the picture of snow geese I present in this book.

Of course the sketches – the résumés – I've presented in this chapter

Barbara Ganter with M-60
on Wrangel Island

are *only* introductions, for even after a couple of years or more of dealing with, say, a Sean Boyd, on a somewhat regular basis, I know only that part of him he chose to show me – the curious, seemingly dedicated and ambitious scientist, whose good nature and sense of duty impelled him to cooperate with me more or less patiently even though he was clearly squeezed for time.

It is important to recognize that his view, and the views of the other scientists here have been informed by the views of the scientists who preceded them, some of whom I refer to in this text, others whom I haven't even heard of. If a scientist's perception of snow geese could be dissected, I imagine it would resemble a slice of bristlecone pine, the oldest known tree on earth: I imagine ring after ring, representing genetic influences, including personality, family influences, cultural influences, educational influences. Does a population biologist who is an optimist view changes in snow goose numbers differently than one who is a pessimist?

I think of the lists of questions I suggest my students ask themselves about the fictional characters they are trying to create. How much do you need to know about your characters? 'Theoretically,' says the writer and editor Rust Hills, 'a character's characterization needs to be no deeper than necessary to fulfil his function perfectly in a perfectly wrought story.' Theoretically, that is.

# 5

# Data Surfing and
# Other Observational Sports

IF YOU ACCEPT that seeing is active, perception selective and constructive, it follows that different people actually *look* at things differently. I
came to the snow geese as a woman, a mother, a writer of fiction, a
teacher, and the niches I fill in life affected both how I observed and
what I observed. For example, I am someone who dwells in her thoughts,
sometimes ignoring what's around me in favour of what I have created in
my own mind. That's the occupational hazard of a fiction writer. Often,
as I sat in my car, or on a bench or a log, or walked along the paths of the
Reifel Sanctuary, the dykes above the Skagit River, my attention wandered from the scene in front of me to thoughts of how I planned to write
this story. What's more, the wintering period coincides, frustratingly,
with my college term schedule, so over and over again I felt the pressure
that comes from trying to fit too much activity into too little time. From
notes I took during a few seasons, I find the time issue recurring again
and again.

'To become an observer one has to allow time for the road, the traffic, the other thoughts on one's mind to dissipate, and become settled

into the scene. Quiet. Become used to what's there so one can begin to observe changes.'

And, another season. 'I feel pressed for time. Frustrated. To see one must look for a long time and I'm always rushing these days. Annie has a hamster home from school for the spring break and she watches it scurry or sniff around in her hand for long periods of time. Watch like an interested child.'

Still later. 'To look and to see one must have access, TIME. One must organize the information. A natural first seems to be to register amounts. First reaction – look at all those birds! Then to pause to count. Notice one behaviour, survey to see if it is common behaviour. How long is long enough to watch? Now I think it is like writing. One must pay attention and go slowly, be patient.'

Although the scientists I talked to represent two generations, both genders, and four different countries of origin, meaning four different sets of cultural influences, they aim to achieve objectivity through their methods. Every report on a scientific study begins with a section explaining the methods that were used. Readers who believe that the methods are valid are more likely to accept the findings of the study.

In the context of the story I am developing here, methods are important as a way of describing how scientists observe snow geese, which is, of course, in a more formal, more structured way than a non-scientist such as myself. While I might look at a flock of geese flying over and wonder how many there are, scientists formalize the 'I wonder' question into a population survey. The Pacific flyway has been keeping a running count of white geese on the Fraser and Skagit for fifty years, and since 1987 Sean Boyd is the man who has done the counting. There are various ways of estimating how many geese make up a flock – such as noting the space ten to twenty occupy, then adding by spaces of the same size – but Sean actually counts snow geese one by one, working from aerial photographs that he pieces together to form a picture of both deltas.

One day I watched Sean's technician Saul Schneider working on aerial photographs of a farm field on Westham Island that Sean had photographed on 8 December using a telephoto lens on a camera with auto advance. The result reminds me of an old sepia print, brown background with creamy grains of sand on it, except that the grains are actually geese seen from 400 metres. Saul uses a device he and Sean invented, a ballpoint pen whose cartridge they replaced with a compass point. He peers through a microscope to better see detail on the photo and pricks each dot with the pen point; the tallier attached to the pen keeps track. It's a very fussy business; he can work for only about two hours at a time before his eyes want a rest. At 3,000 per hour, it will take two or three days to make it all the way through a single series of aerial photographs. But there are fewer photographs to work with now. From 1987 to 1991, Sean spent 250 hours in the air, taking off in a small plane from Richmond every two weeks, then working his way south along the Fraser Delta, then down to the Skagit. The information he gleaned from these surveys revealed how snow geese use the two deltas, so that the past few years he has been making only five three-hour flights a season, for fifteen hours worth of photos. Just enough to keep track of how many geese are here.

The count on 2 November 1994 – the first aerial survey of the season – totalled 48,158 birds, 33,000 on the Fraser, 15,000 on the Skagit.

Besides illustrating the size of the wintering population, the aerial surveys also reveal the year's breeding success because the grey-feathered juveniles show up as darker spots on the photos. This is one of the ways Sean was able to determine that juveniles made up 5 per cent of the population in 1994. He can also estimate family sizes by analyzing the photos carefully, but another way of doing this is to simply watch the flocks in the field.

Saul Schneider explains why and how this is done. 'We're trying to get an idea of the numbers of adults and juveniles per family and look at changes in family size between fall and spring. When the flocks are feeding and resting on the fields, or the marsh, they take off in small family groups and we can determine which are adults and which are juveniles.' He records his observations on tape as he watches, noting that, for example, a group of four contains two adults, two juveniles; a group of three, two adults and one juvenile. 'Sometimes there are no adults, just juveniles. We try to get a large sample of family sizes, and then we'll do some statistics on them, get an idea of what changes take place over the winter.'

As he watches the geese, he picks up other information, not necessarily on the agenda, but interesting all the same. For example, he's noticed that when they are grazing, a certain number of geese keep their heads straight up. 'They're most likely keeping watch for eagles. When an eagle goes over, they all lift up and circle around, then land in the same field again. Some act as sentinels. And I often see a bit of antagonistic behaviour, a goose driving other individuals away from a patch he's feeding on, or charging with their head down. In the fall I noticed adults drive other juveniles away. This spring I saw juveniles driving each other, and even some adults. The juveniles get bigger and more aggressive over the winter, that was my impression. In the fall they can easily be pushed off.'

Saul has also noticed that if all the birds are out in the marsh, a few will fly around looking for different places to feed. 'And I think there's a small fragment that does the exploring. I don't know if this is intentional or if they just happen to find some yummy patch and others notice and join them.'

These impressions, the 'hunches' Saul gets as an individual, while he is collecting so-called hard data, are undoubtedly the result of the time he spends in the field. Other people, who are not scientifically trained, but who also spend considerable time watching geese, notice the behaviour of individual geese too. But the scientific way to actually monitor individual behaviour is to observe individual birds. This would be next to impossible without identifying them somehow, and the traditional way of identifying birds since the nineteenth century has been to place numbered bands on them. 'Neck-collared birds tell us family structure,' says Fred Cooke, 'the extent of family break up, the extent to which they choose mates from different populations, and where they choose them.'

And they help Sean Boyd determine which geese make up the flocks he surveys from the air. Though he counted over 48,000 geese on the Fraser and Skagit Deltas in early November 1994, field surveys of collar numbers around the same time revealed that a portion of these birds were simply passing through on their way to California.

Two kinds of identifiers are now used on Wrangel birds, plastic leg or tarsus bands, and the easier-to-read-from-a-distance neck collars, which – on the Pacific flyway – are red plastic and weigh about five grams, and are printed with alphanumeric codes in white. Russian and Japanese scientists have used yellow and green collars. Some neck collars are fitted with minute radio transmitters that weigh forty-five grams, and a few of these have satellite radio capability.

Though the Russians began studying the snow goose colony on Wrangel in the 1940s, international agreements with Canada and the US have provided for cooperative funding of snow goose research so that the banding and collaring program has been stepped up.

In 1994, Barbara Ganter was one of three North American scientists who worked at the banding camp on Wrangel, a simple tarpaper hut with a couple of outbuildings and tent accommodation for visitors. From a rudimentary observation tower on top of the cabin, Barbara, Don Kraege from Washington State, and Mike Samuel from the US Department of the Interior's Wildlife Health Center in Madison, Wisconsin, who was there to take blood samples from the geese that winter in California, would watch the horizon for sign of Vasily Baranyuk. Baranyuk, the long-time head of snow goose research on Wrangel, would motor across the tundra each day in his Argo, an all-terrain vehicle, until he found a flock feeding around one of the lakes, then herd it back to camp where nets were set up to hold the geese. When they saw him coming, the team members would take their positions at the net.

Barbara's slides of the operation reveal a flock of geese breasting a rise, necks sticking up, running on their dusky-red webbed feet, squawking, moving as a herd towards the corral of nets. 'The little dark things are the babies,' she explains. The adult geese, though, who have moulted their black primary feathers, are all white.

When they get to camp, the still yellow-headed goslings are placed in a separate enclosure so as not to have to endure any more stress, and then the work begins. Barbara describes a typical session. 'It took us a while to get into a routine, because we had not worked together like this before and we were doing so many things to each bird. We were cleaning them, we were swabbing them, we were vaccinating them, we were putting three sorts of markers on them. On the first day we did fifteen birds an hour. We had about forty birds so it took two and a half hours. Later on we had bigger flocks, of about 170 birds, and we were much faster. We could do forty birds an hour. In general everybody had one job. Vasily would weigh them and record their face stain, and sex them and sometimes measure them. And Natalya, Vasily's wife, recorded. Mike and Sergei (geneticist Sergei Kusnetsov) were only there for the bleeding part, so they would take the blood samples and vaccinate. And Don and I would put bands on them and sometimes measure them. But because the collar glue had to dry, putting bands on took a long time.

'You put the neck collar on, and the overlapping bits are glued. While the glue is drying you put a clothes pin on to hold it, and depending on the how cold it is it can take anywhere from ten seconds to half a minute to dry. The glue is a solvent that melts plastic. So you have to hold the goose for a while. Look at my coat: it was completely messed up with collar glue, because on cold days I had to wear everything I had.' Her coat was not the only thing damaged. Though the goose in the picture she shows appears relatively happy to be sitting on her lap, geese can and do bite, and Barbara's fingers have bled from goose-inflicted wounds.

And then, after all this, many neck collars are lost within a year. 'Initially birds really fight them, it takes them a while to get used to them. Many collars last for many years, but it seems in the first year, there's a high rate of collar loss. So we also put leg bands on some birds to find out how many neck collars have been lost.' Normally some of the goslings would also receive leg bands, but in 1994 the goslings that the team encountered appeared so weak, they left them alone.

Still photographs cannot capture the sound of a flock corralled, or the organized chaos of the work. Against a background of brown and grey – summer on Wrangel Island – on what Barbara describes as a very cold day in late July, Natalya is sitting on a box, dressed in a reindeer-hide coat, cigarette dangling from her lips as she writes on a clipboard. 'She was really good. People were yelling at her in Russian, other people were yelling at her in English, and she was just recording it all. Recording is a really important job when you're banding geese. And it's not a very pleasant job.'

A hands-on person, who says she's never happier than when she's sitting in a field of geese, Barbara's favourite part of the work was banding and determining the sex. To sex them, 'You have to open up the cloaca of the bird. Most birds don't have penises. But geese and ducks do so when you open the cloaca, the penis pops out with males, and it doesn't with females. That's how you know. With these birds, because they had just bred, you could tell which were which because females, when they incubate eggs, develop a big brood patch; they lose feathers on their belly so that the warmth of the body goes right into the eggs.'

She clicks the control mechanism on her slide carousel and an image of geese running down to the river appears on the screen. It's sunset, about 11 PM 'Here they are leaving again. When we let them go we try to keep the flock together so families stay together. We try to herd them down to the river so they don't panic so much.'

There were plans to place collars on 2,000 snow geese the summer of '94, but weather had so interfered with the breeding season, the group Barbara worked with was able to put collars on only 750. Fifty-nine of these had radios attached. These were meant specifically for the California birds, because there is concern on the Pacific flyway about a persistent decrease in the proportion of the Wrangel nesters that winters in California. Scientists fear the decline might be the result of avian cholera.

US Fish and Wildlife Service biologist John Takakawa is able to track the birds wearing collars with satellite radios all the way down from Wrangel Island, which has supplied a tremendous amount of information about where the geese stop on their migration route. Birds with conventional radio collars are monitored both on the Fraser and Skagit Deltas and in California. Monitoring helps researchers discover the fate of birds that were inoculated against avian cholera.

Though only California wintering snow geese were supposed to receive radio collars, some ended up on birds that normally winter on the Fraser and Skagit. In the chaos of banding, scientists have only minutes to decide which are northern birds and which are southern. They do this on the basis of facial stain, as described in Chapter 3, but some stains are difficult to interpret and thus some Fraser-Skagit birds mistakenly received radio collars too.

As part of their winter monitoring work, Sean Boyd's technicians Saul Schneider and Barbara Pohl record the alphanumeric code printed in white on the red plastic collars, as do Mike Davison and his assistants on the Skagit, and four to five other people, including John Ireland, manager of the Reifel Sanctuary, and his wife, Mary Taitt. On the dykes above Skagit Bay I met a couple of women from Seattle, turkey vulture enthusiasts, who were copying down numbers they observed in response to a request they had read in a newspaper. And people like Roy Phillips, the retired parks board employee I met in October on Westham Island, do the same. Amateur ornithologists have always contributed information to scientists, casual observations as well as 'harder' data, such as collar numbers. But Sean is careful of the numbers he gets because information can be extremely misleading. 'The worst thing is to have faulty information. A bird might have been here for years, and then we get a number mistakenly reported in California that would lead us to think the flock is breaking up.'

One day in early December, the 6th, just after a night of light snowfall, I went with the two German Barbaras and their telescopes to watch geese and collect data, as they put it, specifically collar numbers. After an hour or so looking for the main flock, we finally found them in a field the Wildlife Service maps designate 45/1A, but which is known to local farmers and hunters as Roddy Swenson's place. It was at first hard to see the geese because they were mixed with clumps of snow. The two Barbaras set up their scopes, and we stood there for the next almost two hours, Barbara Ganter scanning from left to right and back again, slowly, looking for collars. She would find a collar, record it in her book, note which ones she had seen before. She had recently returned from Queen's University where she had successfully defended her dissertation and was preparing to return to Germany after Christmas. But she wanted to personally spot and record every bird whose neck she had encircled with a red plastic collar four months earlier on Wrangel Island. That morning she found forty-eight.

Barbara Pohl was counting proportions of collared to non-collared birds. She had one counter in her left hand, one in her right. On the road the snow had melted, but it still clung to the ochre weeds in the ditch between us and the fields. A sharp damp wind blew up from the marsh. I passed my small thermos of coffee. We all added another layer of clothes.

Sean Boyd (on the right) and John Takakawa
with banded goose on Wrangel Island

It was an entirely different sort of day two months later when I drove with Mike Davison around Skagit Bay. The sun flooded through the window of his Washington State vehicle, glinting off the metal big-horned sheep that ornaments the hood. Blue sky, blue mountains, early shoots of sedge and bulrush making a green fuzz on the brown marshes, acres of daffodils beginning to bud. He stopped at various points along Skagit Bay and Port Susan Bay, turned on his receiver and scanned for several seconds, listening for the chirp that indicated the presence of a snow goose with a radio collar, then copied down the frequency number on a sheet on his clipboard.

During hunting season on the Skagit, the geese had stayed far out in the bay, making it difficult to monitor radios or record collar numbers, even to see them well. The geese have to cooperate if one is to undertake any kind of observation, formal or informal. You have to be able to find them, and they have to stay put long enough to count or photograph, sketch, study.

This is why so much scientific work has been concentrated on the nesting grounds. The Russians had been working under difficult conditions for years before Graham Cooch even knew the colony on Wrangel existed, but when he finally did learn of it, even though it was the height of the Cold War, he was able to make contact with various Soviet scientists and begin the talks that led to the agreements under which snow goose research is conducted today. Conditions are better in many respects, but the Russian economic crisis has meant a lack of fuel, money, and helicopter time. 'Because their helicopters are about twenty times as expensive,' says Sean Boyd, 'we can't get them to do aerial surveys for us any more. It takes $25,000 to fly for a couple of hours whereas a couple of years ago it took $5,000. But they seem committed to it.'

In fact all the North American scientists I spoke to expressed admiration for the work the Russians have been able to accomplish with their limited means. According to Evan Cooch, 'The Russian scientists deserve an incredible amount of credit for working under the conditions they do. But their numbers are probably a hundredth of what we have from La Pérouse Bay. My dad got us involved with the Russians and on one of his first visits he noticed that they had one photocopier for 1,000 people, that type of thing. They had people whose job it was to transcribe copies of things. By hand! It's a different situation, financially restricted. Even things like computers. They got their computers only a few years ago. Fred got $50,000 to $100,000 for twenty-five years. In one year we get more money than they'll get in five. It's a precarious situation, politically, financially. We're attempting to do all sorts of things but we're all cognizant of the fact that whatever money we put in is a risk.

'So what we're trying to do now is add to what they can do. Put money into marking as many birds as they can. By marking up there [on Wrangel] and marking on Banks Island, we hope to tell a lot more of the story of how many move from where to where and why.' Evan is interested in the movements because within the Wrangel population, particularly the portion of the population that winters in California, there is an exchange that occurs with geese that nest on Banks Island in Canada's western Arctic, the population exchange I referred to in Chapter 4, which makes for such complex management challenges.

Fred Cooke describes how it happens that geese from Wrangel and Banks Island can get mixed up. 'Some birds that winter here move on to California. Some birds go on to California by mistake because they're carried along by a group they're travelling with. Then there are all the events at the nesting site: a female from California can slip one of her eggs into the nest of a bird destined for here, so it changes the genes of that population. The other way is that they do have extra-pair copulation. A male from one population can copulate with a female from another population, again exchanging the genes. Then there is fostering, families getting muddled up during the early brood-rearing period, so that a bird destined for California gets into the family of a bird from the Fraser population.

'And they can exchange genes in the south, if a bird from one population gets into the population of another and then chooses a mate from that population. If you just have one population, if you know the birth rate, and the death rate, you can predict from facts you already have. And you can model it: what if there were a vast increase in the frequency of foxes? What if hunting pressure increased? What if there were an environmental disaster? What if there were three or five bad years in a row and there's no production? You can model all that and therefore get some idea of what to do if the population is in trouble. That's if you have a single population.

'But if we have a population that has three bits to it and there's a certain amount of exchange, a disastrous event in California would have no effect on this population if there's no gene exchange, but could have a massive effect if there was a lot of gene exchange. So that's why we have to deal with it as a three-population problem, and that's what people haven't done in the past.

'That's what we can do. Our expertise with population modelling at La Pérouse Bay allows us to do population modelling here. And there's a very good data base here. The Russians have been collecting information for a long time, Sean has been. There's not very much on California and very little at all on Banks Island. But we'll use existing data, and concentrate on gene-flow questions.'

To monitor the exchange, scientists need to follow the movements of individual birds, which is why more and more birds will be banded and collared on Wrangel as long as there are funds to do so. But to gain information about variations in productivity, and to learn specifically how and where the snow geese nest, scientists must work directly on the nesting grounds. Some idea of the intensity of nesting-ground studies is represented in the package of instructions handed out each summer to fieldworkers at La Pérouse Bay, Manitoba, site of the very long-term study conducted by Fred Cooke. When he first established his camp there, in 1968, he was accompanied by only one graduate student, but over the course of the next twenty-five years many more biologists, graduate students, and volunteers became involved. Because the answers he was looking for might be biased by individual views if people went about collecting data in their own ways, methods had to be standardized. So Fred and his colleagues compiled a brochure that explained exactly how to proceed during the initiation, hatch, and banding sections of observation. The following sets out what observers should do during the initiation period:

'Initiation, as the name suggests, is the start of our research. The initiation period is the time from the finding of the first nest to the day the last egg is laid. The start of initiation coincides with the appearance of bare ground which is usually in the middle of the spring melt. There is still a lot of snow and deep water so walking can be quite interesting. Each person working on the Snow Goose Project is assigned an area to search intensively for nests. The size of the "intensive area" covered is governed by the amount of time available. Each nest that is found receives a letter and a number, for example A21 would be the 21st nest found in the A area. Each new egg in the nest is numbered with a felt marker and weighed; each nest is visited every day until incubation is started. Ideally each nest should be found at the one egg stage and monitored through to hatch. These would be considered the best possible intensive nests. As the nest density in an area increases it may become impossible to cover and search effectively the entire area that was searched on the first day and revisit all the nests found previously. At this stage the field worker may have to reduce the size of the intensive area so that the entire area can be searched for new nests every day and at the same time check all the previously marked nests. It is very important not to try to cover an area which contains so many previously marked nests that you have no time to search for new nests. This would result in a biased sample containing only early nests and this *must* be avoided. As soon as possible each worker should try to map the nests on the large scale aerial photographs. This requires practice and the easiest way to learn is by accompanying an experienced mapper in the field for a short while. Once you have identified a couple of landmarks in the area the rest of the mapping may become easier. As each nest approaches the start of incubation you will find the pair will stay much closer to the nest. At this time it is often possible to record the phenotype, sex and bands (if any) of the parents (see section title "Notes on Band Reading, etc."). All of the information for a single nest can be recorded on half a page in your field book. In order to illustrate the code we use I will now present a couple of hypothetical examples.

|  |  | L | R |
|---|---|---|---|
| A003 | (1) | NB | NB |
| willow, .5m, wet | (1) | NB | NB |

30/5 $1^{120g}$ \ 31/5 1 \ 1/6 1, $2^{141g}$ \ 2/6 1 M, 2 cold, \
3/6 2 PG \ 4/6 \ 5/6 T \

Translation: Nest A003 was found on May 30 with one egg that weighed 120g. The predominant vegetation within 1 m radius of the nest was willow, the nearest snow was .5 m away and the bottom of the nest cup was wet. On May 31st there was no change in status. June 1st a second egg was in the nest and it weighed 141 grams. June 2nd egg 1 was missing and egg 2 was cold. June 3rd egg 2 was preyed upon by a gull and on the 4th and 5th nothing was added to the nest. The nest was Terminated "T" on June 5th because there had been at least 3 days during which no new eggs had been added to the nest. It is assumed that the pair had abandoned this nest. The final thing to note is the pair was seen at this nest; they were both white (designated by the 1) and had no band on either leg. (NB). You should always record the bands you observed even if you only get a partial observation. If you never see one of the legs clearly enough to be sure, record this as "leg not seen." You should also record the date(s) on which you identified the phenotype(s) of the attendant pair. The upper record is of the female and the lower is of the male. You will be shown how to recognize the two sexes.'

The code letters and numbers and symbols are explained, and then a note is added. 'Any fates not included in the above list, such as "cracked egg" or "preyed upon by polar bear" should be *written out in full.*'

Approximately twelve areas (about 3,000 nests) in total are monitored at hatch in most years.

At hatch, the monitoring begins with observations of eggs. Student and volunteer observers try for a minimum of disturbance, try not to change the normal progression of things, to the extent that they have to recognize and act on the difference between an emotional and scientific response to various situations. That the two can be mixed up is reflected in a piece of instruction regarding the last egg to hatch, which the mother often leaves behind as she follows the first hatchlings off the nest to feed. When the last egg does finally hatch, and a lone gosling emerges, it is a doomed orphan because workers are instructed to ignore abandoned goslings *'no matter how irresistible they look.'*

The same methods were, and still are, employed year after year, says Evan Cooch, who participated in the summer fieldwork for three or four full seasons and bits of six others. 'It's enjoyable to actually go up there and physically do it. But most of the work up there in my opinion could be and more or less is done by technicians and grad students. It's not mentally taxing, it's physically taxing. You're up there three or four months. You're out there every day, nine or ten hours a day, running around, collecting data. Measuring eggs, how many eggs, finding nests and looking at birds. It takes a lot of time.

'The next level is generally putting it in computer files. Now the question is, what data do you collect? The flippant answer is, everything you possibly can. The practical answer is, what you can feasibly do within time or money constraints. So there's nothing really magical about counting eggs or seeing which birds are in a field. And there's nothing magical about taking that from a piece of paper and putting it onto a computer file.' He swivels around on his chair and clicks on his computer.

'For instance I've got a file here, nest data from 1973 to 1992, each of these numbers represents something. We've got 42,000 nests' worth of data.

'Then the question is, and that's what you get from your biological and scientific intuition, what phenomena do you believe might occur in your population? Do young birds and old birds basically do the same things every year? Is age a factor? Is it a factor for what? You then analyze data using applied statistical tools to answer that question. But it almost always starts from the question, gee, I wonder if geese do this? And then we say, we've got the data to ask that question. For example, it looks like as the habitat around LPB [La Pérouse Bay] deteriorates, more birds are leaving. I wonder if it's the young birds that leave. There's theoretical information out there in the world that predicts that in a deteriorating environment, young birds should leave. So there's a hypothesis, we can test it. And that happens 50 per cent of the time. We've got the ultimate real world data set and we can say, there's the prediction, let's look at our data and test it.

'The other 50 per cent of the time we say, okay, we know the following factors tend to be important, so let's just look at more things, like egg size. We call it data surfing. Get in and start mucking around. Most of the day-to-day thing becomes a technical exercise in how you extract information from data that exist. To a degree we have to move away from reality.

'But these numbers represent data that were collected in the real world. The real world is complicated and we're now trying to find the pattern, the process under the mess of data. The idea happens very quickly, it takes five minutes. Then it takes fifty-five minutes to figure out, now how am I going to ask that question of the data that we have? And so, in an eight-hour day, maybe an hour is spent thinking about the question, then seven are spent mucking around writing the programs to do it. You get a number or a set of numbers and you say, oh, okay, then you say, well I wonder if ... so you learn a bit, then go on and do another bit.

'What Fred has brought to the project, has always brought, his singular strength, is that he has a very good scientific vision. I mean it's world class. But he doesn't do any of this. He brings in people to do it. Like myself, and David Lank, computer nerds who are also biologists. We'll spend weeks on analysis and come up with something and sit down and tell him, and he'll say, well that doesn't make sense. Or sometimes he'll say, that's great. Because he has this detached view he can sometimes ask these penetrating questions. It's a symbiosis that's been very fruitful.'

For people who associate wildlife biology with long trips to exotic places in the wild, Evan's approach seems extremely detached, but as he explains, 'I look at numbers and computers as the tools that allow me to do the thing I like to do most and that's ask why. I accept, *I believe*, there are fundamental reasons for why things happen. And I get a real chuckle out of figuring out what it is. The computer allows me to ask a question and get an answer and that's psychologically satisfying. For one minute I know something nobody else in the world knows. Maybe it's irrelevant to the rest of the world, but I know something.

'It's a different path to knowledge. The quintessential tribal elder can tell you how many fish are going to come down the stream because he's been looking at them a long time, he just knows. My dad was like that. I don't have that ability. I use different tools to get to it, a way that suits me personally.'

So where does all this data collection and analysis lead? To information that is interesting for its own sake, and to information that tells us something about the world at large. For example, at La Pérouse Bay, when Fred set up his camp in 1968, the nesting colony consisted of about 2,000 snow geese. The colony has since increased tenfold. The result is that individual birds are getting smaller because they are eating their food supply out. The salt marshes the snow geese depend on at La Pérouse Bay are disappearing, which will lead eventually to a classic crash. Fred has surmised that the population increase is related to modern agricultural practices on the wintering grounds.

'The way agriculture is done, people leave about 5 per cent of the crop on the ground.' This is valuable for geese that used to have to depend solely on the marsh sedges of the Gulf Coast. 'So fewer geese are dying over the winter. More go back to the Arctic. The salt marshes in the Arctic are about to disappear. That's something that could affect an entire ecosystem most North Americans are totally unaware of.

'By doing this detailed study we've been able to identify the problems of population growth in geese. Destruction of arctic salt marshes is probably a consequence of human agricultural practice more than anything else. And we've alerted people to the problems of the disappearance of the salt marshes. A lot of other species are affected by the expansion of snow geese. It may be fifty years before salt marshes come back. This led me to the realization that you need careful demographic information to solve a lot of the conservation problems. Those problems could not have been solved by simply looking at the winter population or at the hunting pressure. Only by putting it all together could we identify what the real conservation issues were. That wasn't my objective when I started, but it resolved the question in a way all the wildlife people who had been studying snow goose populations for years never hit on. They were interested in hunting pressure.

'This was one of the strengths of Graham Cooch and Hugh Boyd. They believed that you got the best answers to wildlife management questions by looking at the basic ecology of the species. In my case I hadn't even thought of the conservation aspect of it, I was interested in the geese. As it turned out, our project was the one that discovered the link between food, salt marshes, and population changes.'

# 6

# *The Artist's Perspective*

Scientists are required to erase themselves from the pictures they construct; at the scene of the perfect inquiry there should be no telltale fingerprints, no tracks leading back to the inquirer. Ideal science is a text without an author. The conventional way of putting this, of course, is that scientists aspire to objectivity. They strive to produce descriptions of reality that are independent of the biases of the individual scientist – to reveal the world as it is in itself. And the method works remarkably well. No one familiar with the accomplishments of twentieth-century science can deny its effectiveness.

Problems arise, however, when the scientific method is considered intrinsically superior to all other ways of coming to grips with the world, because when elevated to that status the sleight of hand that makes it possible becomes a liability. And what sleight of hand is that? Disguising the origins of the inquiry itself: pretending that it is something other than a uniquely human enterprise, that its very existence is not a manifestation of interest and values reflecting rarely examined prepossessions about the nature of truth and the nature of man (Edward Dobb, *Harper's Magazine*, February 1995).

WHILE SCIENTISTS are committed to objectivity, artists go to the other extreme, presenting views unabashedly their own. Subjectivity is intrinsic to art. Yet it is often the artist's vision that creates the most indelible impressions of reality. In the case of snow geese, it was the description of the lone snow goose in Paul Gallico's story that first caught my imagination:

'One November afternoon, three years after Rhayader had come to the Great Marsh, a child approached the lighthouse studio by means of the sea wall. In her arms she carried a burden ... a large white bird, and it was quite still. There were stains of blood on its whiteness and on her kirtle where she had held it to her ...

'The bird fluttered. With his good hand Rhayader spread one of its immense white pinions. The end was beautifully tipped with black. "It's a snow goose, from Canada, but how in all heaven came it here?"

'... The bird was a young one, no more than a year old. She was born in a northern land far, far across the seas, a land belonging to England. Flying to the south to escape the snow and ice and bitter cold a great storm had seized her and whirled and buffeted her about. It was a truly terrible storm, stronger than her great wings, stronger than anything. For days and nights there was nothing she could do but fly before it. When finally it had blown itself out and her sure instincts took her south again, she was over a different land and surrounded by strange birds that she had never seen before. At last, exhausted by her ordeal, she had sunk to rest in a friendly green marsh, only to be met by the blast from the hunter's gun.

'"A bitter reception for a visiting princess," concluded Rhayader. "We will call her 'La Princesse Perdue,' the Lost Princess. And in a few days she will be feeling much better. See?" He reached into his pocket and produced a handful of grain. The snow goose opened its round yellow eyes and nibbled at it ...

'In mid-October the miracle occurred. Rhayader was in his enclosure, feeding his birds. A gray northeast wind was blowing and the land was sighing beneath the incoming tide. Above the sea and wind noises he heard a clear, high note. He turned his eyes upward to the evening sky in time to see first an infinite speck, then a black-and-white pinioned dream that circled the lighthouse once, and finally a reality that dropped to earth in the pen and came waddling forth importantly to be fed, as though she had never been away. It was the snow goose.'

And later, when Rhayader, a cripple, an artist, and a lover of wild birds, sails off to Dunkirk to help rescue fleeing soldiers, '"A goose, it was. Jock, 'ere, seed it same as me. It come flying' down outa the muck an' stink an' smoke of Dunkirk that was over'ead. It was white, wiv black on its wings, an' it circles us like a bloomin' dive bomber. Jock, 'ere, sez: 'We're done for. It's the hangel of death a-come for us."'

A lost princess, a black-and-white pinioned dream, the angel of death. These were some of the views of snow geese the literary artist Paul Gallico ascribed to the characters in his story. His lyricism had a powerful effect on me long before I ever moved to the west coast of Canada and saw a snow goose for myself. But the images Gallico presented all concerned a single snow goose, and it is so unusual to see a single snow goose, or even two or three together in one place, that I couldn't believe my eyes when I visited the Stanley Park Zoo in Vancouver and saw – thought I saw – a pair of snow geese swimming around with the muscovy ducks and domestic geese, the mallards and wigeon. They were indeed snow geese, however; the children's zoo director, Allan Reagan, confirmed it. Two males remaining from a group of ten or twelve, most of which had died from duck enteritis several years ago. Not prevented in any way from flying off, the two snow geese stay, apparently having accepted this immense and often busy urban park as their year-round home.

So what do artists see when they encounter a field of snow geese, I wondered. The individual? The flock? As with scientists, the answer to that question is bound to be as diverse as the many painters and more photographers snow geese have inspired. But in our time, the painter whose depictions of nature in art are known just about universally is Robert Bateman. Coincidentally, in the family tree I was constructing of snow goose observers I found yet another connection. Fred Cooke and Bob Bateman had met in Ontario, when both served on the executive of the Federation of Ontario Naturalists, and Bateman had actually spent a couple of weeks at Fred's camp in La Pérouse Bay, observing the hatch and putting web tags on chicks.

'I've always liked Fred because of his lively mind and sense of humour and enthusiasm, and I like English types, I'm a bit of an anglophile. So he was telling me about his work and hinting that he sometimes took volunteers up to La Pérouse Bay to help with the banding of the hatch, or the banding after the moult. I said gee, that's the kind of thing I'd really enjoy getting in on. I said I'd rather go at hatch because it's springtime and all the other birds are around, it's full of life, so that's what I did. I came in after the eggs were counted, the nests marked. I would go around in my territory every day and see if the eggs were hatching. And if they were, we'd put little web tags on the chicks.'

His hands were too full to do any painting during that time, but he did some sketches, and a study in oils for Fred. Some years later, in 1983, he completed another painting of snow geese, a wintering flock flying above a field in California. In his comments on that painting he mentions having been infatuated with the restlessness of the geese, the clamour of their conversation, the sense of air and space and sky 'heightened immeasurably by the skeins of geese cutting pathways through the air at different heights and in different directions. I thought of dancers swirling veils in arcs through the air.'

But there was a difference between watching the big wintering flocks and his experience at La Pérouse Bay. 'I've seen the big wintering populations, I mean it's the opposite end of the scale, right? At La Pérouse Bay they were on their own territories with their own characteristics, the other was just this massive display, a big spectacle. La Pérouse Bay was much more intimate and it became like a family affair; I was involved with these families. And of course we had to mark whether they were "sticky" geese or not: some geese were called sticky because they were brave and stuck around the nest when humans approached. Fred wanted to see if that quality was inherited. Other geese would sneak away. So I had to go back day after day for the whole time I was there and it was just like I had gone back to the Garden of Eden. I was in this bounteous world, eider ducks and oldsquaws and ptarmigan, and beautiful weather like this and sloshing in our hip waders through champagne coloured shallow water, stepping on these little knolls that were just like bouquets of flowers and little bonsai trees. And it was oh ... it was just wonderful, and being tilted toward the geese and seeing all these little goslings coming out and the wonderful light on them, and the sticky geese when they tried to scare us away. Just for the senses alone.'

Besides the experience they shared at La Pérouse Bay, something else connects Fred Cooke and Bob Bateman, a quality of mind. When I mentioned to the scientist husband of a close friend of mine that I was going to be meeting with Fred Cooke, he remarked that although he knew Fred as a very busy man, having served on various committees with him, he had always admired Fred's ability to focus the whole of his concentration on whatever he was doing at the time. 'You'll find he gives you his total attention.'

Bateman is a busy man too, unable to resist the many interesting opportunities that come his way. 'I'm like a kid in a candy store,' he says. 'I find it hard to say no.' And so he travels several months of the year, attends this function, donates that painting for a worthy cause, serves as honorary chairman of this or that committee, all the while producing the paintings and prints that have opened these doors for him, and spending time with his wife Birgit and their two children, plus the three from his first marriage. How does he do it?

'The whole idea of living in the moment, something that's in all religions – now is the only thing that really exists – I use that all the time. I use that especially in my career, because when you're dealing with 200 people in line, waiting to get their book signed, you can take the attitude that it's a chore – oh how am I ever going to get to the end of the line? But I go into a zenlike state and I only live for now and I absolutely have a ball on these things. Like you, right now, are the only person in the world. I'm there to get enjoyment from the person in line. It's a great honour that they've done that, maybe I can learn something from them. So I take my time and I talk to them. And they're the only one in the world. Because I'm spoiled with so many treats and stimuli, I have to shake myself and take hold of myself and say be focused on where you are.'

Early on in the writing of this book I considered trying to characterize all the people I met as various species of birds. I thought of the marks of recognition, the little arrows Roger Tory Peterson would use to point to features a birder should look for in the field. There definitely seemed to be recognizable differences in the scientists I talked to, for example. As for the species Bateman, a birder might first notice his call, characterized by a ringing open laughter, similar to the exuberant chirp of the red-winged blackbird.

Snow geese fill the fields of Westham Island and the skies above.

In his sixties now, Bateman still exudes a boyish enthusiasm. And though he doesn't think of himself as an emotional person, his spirits rise and fall visibly during the course of our conversation and he comes close to meltdown at 2 PM, which is the time he retires for his daily nap.

When I wrote to ask him if he would talk to me about snow geese, I had fantasized us sitting along the road or on a bench in the sanctuary, me with my notebook and tape recorder, he with his sketchbook, the snow geese feeding, chattering, flying up, settling down. But Bateman's busy schedule does not allow for casual trips across the Strait of Georgia from his Salt Spring Island home to the delta of the Fraser River. The day we spoke on the phone, in June, he could see only one or two days open on his calendar for the rest of the year, and these were in July when the geese were still high in the Arctic, unable to fly because they had moulted their flight feathers.

So we meet on the patio of his wife's childhood home in West Vancouver, where the scene before us consists of ships at sea, the sky-scrapers of downtown Vancouver, the Lions Gate bridge, Stanley Park. A clear summer morning, temperature rising. Following his example, I leave my sandals on the door mat and follow him into the living room, where he spreads an old drape over the carpet, ties an apron around himself, and places on his easel an unfinished painting of a timber wolf, a Mexican subspecies of the timber wolf. For the next few hours we will talk about nature and art, perception, as he continues to work on the painting, something that's possible, he says, because he uses a different part of his brain to paint.

This northern Mexican wolf is an endangered species, and Bateman's composition has it in the background, half hidden by brush, disappearing into the shadows. As morning turns to afternoon he lightens the foreground bushes almost to white, which has the effect of deepening the shadows into which the wolf is disappearing. There's clearly a message here from Robert Bateman the conservationist, whose recent works feature whales caught in fish nets, seabirds blackened with oil, clear-cut forests.

With the same naïveté that led me to believe there might be a sci-entific view, in the singular, I ask Bateman, who is clearly a naturalist, conservationist, and artist, which perceiver comes into play first when he looks at a scene. I had read the following passage in *The World of Robert Bateman*: 'I look at it first as a naturalist and start identifying the different species and looking for what is new and exciting. It's at a later state that the artist in me comes into play and then the two move along together.' Like a concerto for oboe and flute, I thought, imagining one instrument introducing the theme, another embellishing. But of course it's not this methodical.

'The big start of making me the way I am was my mother sending me down to the Royal Ontario Museum Junior Field Naturalists Club. There I started to see adults who were crazy about nature because a lot of museum staff worked as volunteers with the kids, even after hours, and certain of us enthusiastic kids became groupies of museum people. Nine to eleven was the Junior Field Naturalists Club and most of the rest of the kids would go home after that, but we'd sneak up behind the scenes and hang out in the offices of ornithology. We grew up and became teenagers and we still hung out with these museum people.

'My dad is an engineer and my mom was just a classic United Church lady, but very interested in everything in the world and self-improvement, taking courses, doing volunteer work. Being involved. I think I was eight when she sent me down there. Later in life I led my mom and to a lesser extent my dad to an interest, but I was always ahead of them as a kid whether it was art or nature.'

So was it art that came first, or nature? Or can he recall the view he first turned on the world? He doesn't hesitate to say it was art. 'The best evidence I have comes from when I was on a book tour much later in life. A nice white-haired lady came and stood in line, and she was my Sunday school kindergarten teacher in the United Church. When the sermon started they used to send all the kids down to Sunday school rooms, and she said she always remembered little Bobby Bateman, at the age of five, was the best at doing art. The other kids would scribble and I'd do careful flowers. And what was amazing to me was how she remembered little Bobby Bateman. I mean Robert Bateman uh ... When did I start to get known? I'd be at least forty-five before I was known. So little Bobby Bateman was five. Robert Bateman was forty-five. How did she remember that character for forty years? That's what I don't get.

'At any rate, I always was the best kid in class in art. So you can't say nature came first. What I took at the Junior Field Naturalists was bird carving, trying to make the birds look as realistic as possible.'

But when it came time to choose a major in college, he did not choose art or biology because, he says, he did not have the math for biology and he never thought he needed courses in painting to learn how to paint. 'I always assumed, and I still feel this way, that you're born an artist and you're going to be an artist and the courses you take are really irrelevant to whether you're going to be an artist or not.'

Instead he majored in geography, partly to get free trips into the bush, but also because, on the practical side, it would allow him to work as a high school teacher. And sure enough, the topographical maps he was exposed to in geography classes fed both his interest in nature and his growth as an artist. 'There would be a lake that I'd seen on a topographical map way in the bush behind our cottage, that I'd never been to. So I'd say today I'm going to take my lunch and my sketchbook and I'm going to try and find that lake. And I'd take a compass and head off, and read the map etcetera, etcetera ... Turn to the right.' His voice level drops as he goes into his mind to picture the scene. 'Then I'd find the lake. Bursting through the trees, all excited. And I'd just feel terrific, and stimulated and excited that I made some kind of discovery.

'I might sketch it or I might be running late. As happened in one case, there was a heron colony in the middle of the lake. That would be even more exciting, so I'd plan to come back again, maybe bring a little rubber raft so I could get over to the heron colony.

'A bit later in life I was one of the ones who would take part in the breeding bird census. I was asked to do it because I had skills as a birder. I knew my bird calls and so on. It was really neat to feel I was involved in something like that. Virtually all my friends were wildlife biologists and I always felt they were, or used to feel anyway, that they were having more meaningful lives than just being a high school teacher because they were out there on the frontiers of science and I was sort of a thwarted scientist because I didn't have the math. I headed off towards teaching and they did research.'

But he would take advantage of any opportunity he could to work as an amateur scientist, which is why he jumped at the chance to join Fred Cooke at his camp on the snow goose nesting grounds.

'The other thing was, during my twenties I was a mad small-mammal collector. In my reading back then and my association with the museum, I came across people who seemed to have just a fabulous time, like Charles Darwin going to the Galapagos collecting all kinds of stuff and bringing it back to England. A lot of my mentors at the ROM would go to some remote part of northern Ontario, or if they were lucky even Ethiopia or somewhere. And the fellowship of being in the field, making discoveries and collecting, doing drawings and sketches, tracking mammals, and shooting birds and stuffing them. I guess I became a bit of a collector, acquiring a mass of stuff which would then go back to the museum. Literally everywhere I would go I would have a little knapsack of mouse traps.'

Amateur scientist, field naturalist, budding artist, however he defined himself, he was practising methodical scientific observation as a consequence of his obsession with small mammals.

'You go through very standard procedures for museum specimens. You measure the foot, you measure the tail length and you record all that on the labels ... but an incredible thrill to me was one summer I hit the jackpot. I got to spend some of the most fabulous four months of my life around Ungava in arctic Quebec. Four white men and two Eskimos tramping around on ground no human being had ever walked on, according to the Eskimos. They said Eskimos had never been there, to their knowledge. It was virgin territory, really virgin. The previous year I had brought mouse traps with me to Newfoundland. I worked for the Geological Survey of Canada in central Newfoundland and I brought mouse traps and it was a little bit nerve-wracking because I was always a shy guy, especially when I was in my twenties and teens, and I assumed the other guys would think trapping mice was crazy, a bizarre, crazy activity.'

He thought he would wait until the young men he was going to be living with all summer were used to him before he told them what he intended to do with his mouse traps. 'And I'll never forget going off after

Robert Bateman

dinner, heading into the bush to trap my mice ... and one of the guys said, "Where are you going?" And I said, "Just for a walk." "What you got in the knapsack?" "Uh, mousetraps." "What did you say?" "Mousetraps." "How many mousetraps you got there?" "Uh, about a hundred.'" His laughter peals through the bright room above the sea. "'What? I didn't know we had problems with mice.'"

'Anyway, when I got to Ungava, there were four white men and the other guy was also secretly trapping mice for the National Museum. Of all the flukes of fate! He was training to be a doctor, and he got this summer job as a field assistant doing geology research and he had friends at the National Museum and so he was collecting for the National Museum, I was collecting for the ROM. And I was particularly targeting a certain mouse, an extremely rare mouse, a vole with a yellowish nose. Up until that time there were only seventeen in all the museums of the world and I caught seventy that summer. So that was a little feather in my cap, to capture so many of those rare mice.'

It seems clear enough, then, that his artistic vision of wildlife stemmed organically from his interest and experience in the field as a boy and a young man. But he isn't ready to simplify it so. 'That's the meat of what you're all about, isn't it? And that's the most impossible thing to answer. I think I'm a very analytical person. I love analysis. I'm also a chronic planner. I love planning and doing my calendar. But in spite of saying this, that I'm an avid planner and I'm an analytical person, the really important things, the questions you're dealing with are totally intuitive and I'm just as hopeless, or *it's* just as hopeless to express as women's intuition. The real important things, and you already know this, the important decisions people make in life about their mates or where they're going to live or what they're going to do in life are emotional. Very very hard to rationalize. I can make up rational statements for them, but it's after the fact.

'I don't know why I like seeing owls. I can't explain why. I get very excited if I see an owl, it's very exciting. Even now if I would go for a hike I might stop and waste quite a bit of my time and get back late for a phone call because there was a flock of crows when I was having my noontime hike and I'm trying to see what they got. What's it matter what they got? ' The laughter chimes again. 'It doesn't really matter at all what they got. But I just want to see it. And if it's an owl it's a thrill. Why is it a thrill? You could well ask me and I don't know. It just is.'

He cannot rationally explain why he is interested in nature in the first place and why he feels compelled to transform plants and animals that have caught his attention into works of art. But I am beginning to think that his ability to focus, which we talked about when we began, must have sharpened with practice. As he grew older and spent longer periods of time observing, he noticed details. Being an artist, he represented these details in sketches and paintings, and this habit of reconstructing what he had seen honed the breadth of his perceptual abilities. The more he saw, the more he drew what he saw, the more he was able to see.

In *The Zen of Seeing*, the artist Frederick Franck proposes a meditation he calls seeing/drawing as a way of perceiving the natural world. In the workshop where he introduced this exercise, a student remarked: 'I've looked at geraniums all my life but I never saw them until I tried to draw them.'

Bateman has raised the seeing/drawing experience to a skyscraper level of sophistication, which combines myriad sources and techniques. He pulls out a sketchbook he had with him on a recent trip to Spain. 'I use a camera to record and my sketches. I may have taken a snapshot of that, in fact I did. If I ever use it in a painting, I may decide to use a black woodpecker, but the painting may be that,' he says, using his thumb and forefinger to frame a sketch of very tall trees with hills rolling out to the right. 'A tall tall tall skinny thing with the black woodpecker way up there, or maybe way down here. But I'll then use my art intuition, my influence of Japanese art and things I've seen and just my instincts. There are all kinds of possibilities once I have a photograph or a sketch to draw on.

'But I have the action here. And then I do drawings from my own sculpture and then I can put those into paintings.' He leafs through the book until he comes to an image that stops him. 'Now that's an idea I might use for a major painting. It's a Spanish imperial eagle flying along with a hare in its talons. I would never have been able to photograph that. It was coming towards me and I got just a quick glimpse. Very strong.'

He can sketch more quickly than he can set up a photograph, but photographs help to capture detail, and he has referred to as many as twenty or thirty snapshots in a single painting. He uses binoculars and telescopes too. 'All aids help, to make the subject clearer. It's the big advantage we have over Audubon. Those guys who didn't have these viewing aids produced some pretty great stuff. But it does help and I think it's a good idea. There's a couple of other things you may not have thought of. The camera had an extremely important impact on art in that for the first time in history we were able to capture an elusive moment in a split second. And this was a terrific influence on the impressionists, particularly Degas, who became an avid shutterbug and based a lot of his painting on snapshots. Before that, almost all art had been an arrangement. Mona Lisa is an arrangement. She's sitting there with a little bit of landscape completely unconnected with her. The little bit of landscape on this side doesn't seem to be connected to the little bit of landscape on the other side. Whereas the kind of art I like is a slice of life, which Brueghel tried to do. And a few other artists through history. It doesn't have so much to do with detail as a fleeting glimpse of something that's just happened. But it really flourished when the camera came out. So we get these Degas scenes with people turned away and going off the page and faces out of focus, and the centre of the picture may be the floor between two café tables and people way off to one side. And that's the way a casual snapshot sees the world. I love that look. I love that casual snapshot look.'

But he doesn't paint café scenes, or women with mysterious smiles, and because wildlife is his subject, some critics have considered him a trifler in what he calls the 'capital A art world.' Whether Bateman is an artist or an illustrator has been the subject of much discussion, and though he claims he isn't bothered by what people say, my impression was that he feels he has been slighted, unfairly dismissed by some of his peers.

In a television documentary I watched on artists and the environment, Bateman explained that wildlife art has not been in the mainstream of art in our culture, though it has been a part of every other culture. 'After the Renaissance, it was okay to portray domestic animals. But wildlife could not be depicted eye to eye with respect. Art depicting nature was considered to be illustration. There was no easel art depicting wildlife.'

This is partly why, in his development as a painter, he spent a long period on abstract art. But he felt a schism between his 'two selves' when he was an abstract artist, and ultimately turned to realism because he feels that realism is the perfect style for capturing the sanctity of every square inch of the natural world: The particulars. 'Since I became a realist I haven't changed my style very much. I don't feel I have to change. This suits me. This is the way I see the world, and since the world of nature is my subject I haven't even scratched the surface of the visual possibilities. Nature is not only more complicated than we know, nature is more complicated than we *can* know.'

His 'road to Damascus,' as he calls it, his conversion to realism, came when he saw the work of Andrew Wyeth. 'From virtually 1900 to 1962 the subject dropped out more and more from what was accepted as real art until it was just colour field painting in the '6os. Andrew Wyeth, meantime, didn't care, he just painted along, and the gallery in Buffalo had the guts to have a show of Wyeth. And here was an artist who cared about the actual texture of Chadsforth, Pennsylvania: this was an old stone wall, and this was Christina's elbow, and this was the meadow she was lying in and this was her house and all that mattered, and it was okay to care about subject matter. I was by now thirty-two years old, an active naturalist, that was probably about the time I met Fred Cooke. I was serving on all kinds of committees, I was doing leadership-training courses for day camp leaders, Y, boy scouts, I was taking kids out into nature. I was thinking it was really important that kids know the particularity of all the species, that kids know the difference between a sugar maple and a red maple and mountain maple. And yet my art wasn't showing it.

'The variety in nature! Just take a cubic meter of an ordinary vacant lot in Toronto. The action, the visual dynamic, let alone the physical and chemical and biological action of things; just the different light as the sun hits that hunk of weeds and broken glass and all the other things that are going on there, the little things that are struggling to survive and getting killed by pollution and the visual possibilities of that just make the

Another artist's view. Point Roberts artist Evelyn Roth leads a
parade of third-graders from Ladner Elementary School around her
creations of inflatable art, including a forty-five-foot snow goose.

most big complex computer look like a simple little, plain ordinary thing, compared to the dynamics of nature.'

As the psychologist Robert Buckhout pointed out in his 1974 essay on eyewitness testimony in *Scientific American*, 'Perception and memory are decision-making processes affected by the totality of a person's abilities, background, attitudes, motives and beliefs, by the environment and by the way his recollection is eventually tested.'

Bateman unconsciously provides an example of how this works in his case by attempting to summarize how he perceives the natural world: 'It's because of who I am. I pick things because of my entire history. Because of being born in 1930, because of Gordon Paine giving me Cézanne's book on composition, because I got summer jobs in Algonquin Park, and I happened to emulate Tom Thomson and Jackson and the Group of Seven, because I then got turned on and excited by some of the abstract expressionists, Franz Kline and Mark Rothko, Clyfford Still, etcetera. They're all a part of me, and I now see the world through their eyes. Once you've experienced Tom Thomson you can't go to Algonquin Park without seeing it through his eyes. Once you've experienced Andrew Wyeth, you drive through rural eastern countryside in March or November, and you stop and put on the brakes and look at a scene you would have driven right by before if Andrew Wyeth wasn't a part of you. There are all kinds of influences, many of which I don't even know, that are part of my psyche. And they are all sitting there like a whole legion, back in the back of my psyche somewhere and they're making me put on my brakes, metaphorically speaking, and making me go ah-hah, or making me go, oh that's boring, or that has possibilities. And Tom Thomson is working alongside the Sung dynasty and Franz Kline producing my stuff and my intuition, and that's why there will never be – and this isn't an ego statement – why there will never be another Robert Bateman. I can give workshop after workshop after workshop. I'm told there are hundreds of imitators out there and none of them will ever do Robert Batemans. They're trying to. Some of them are consciously trying to because they think it's a stairway to the stars, but they never will because they're not me, they weren't born when I was, they didn't have the same influences.'

Robert Bateman's 'Snow goose landing'

Like the scientists I introduced in Chapter 4, like humans in general, he sees with the eyes of all the people who ever influenced him. When he begins to paint, he transforms what he has seen to serve his artistic vision, the image his mind has assembled from the various pieces of information it has received not only from nature, but from other artists. Although he has been, sometimes scathingly, described as a realist, he manipulates actual scenes to create the illusion of reality. 'I got this way because of Cézanne, or really a book on Cézanne's composition by an author called Loran. He made Cézanne out to be a little machine, very carefully controlled with one little thing fitting into the other little thing. If you took out one part, the whole thing wouldn't work at all. Some artists just despise analysis, but I happen to love it because I'm analytical. So that's happening all the time I'm working, it's part of my psyche this uh ... analysis manipulating. This feeling that art begins where nature ends. To me art is like a little building, a watch or a clock. I use photographs of course, but I put every element where I want it to be, not just where it happens to be in nature, because it works almost mechanistically, the composition works in thrust and tensions and all that.

'Art begins where nature ends. So ... ' he scowls at the painting on the easel. 'This particular scene is California, which I hope – I've done a little bit of research – and I hope it could be Mexico. Mexican wolves don't occur in California. But there are live oaks where they occur.

'If I ever have a conflict between art and nature, I let art win. But I hope I always do it on purpose, that I know that such and such is not scientifically accurate. However, for the purpose of the painting, the painting working, I need to bend the rules. It's what's called artistic licence, I guess. But I try quite hard to reconcile the two.

'So I contrive my paintings as the Japanese do their gardens. I contrive them to look uncontrived. When I feel my painting is looking contrived that's when I'm the most depressed about it, and if it looks as if – to come back to my definition of masterpiece – if it looks as if it just happened, without effort, if there's a freshness to it, then I know I'm being successful, and um ... I mean I could give you some technical ways you can get the sense of reality, and one is don't be a slave to a feather and the fur, but concentrate on light and tone, make atmosphere, the sense of air and space, and those are all just technical things I could teach you in a workshop. But the bottom line comes back to intuition.

'Recently I've been doing more tough art. And part of it is to make a message in my art, but I probably more effectively do that when I give my lectures, which I do all the time. Part of the motivation for my art leaning in this direction is, this is the direction that the art snob, avant-garde, capital A art world also is moving. Ever since I've been eighteen years old I've been conscious of where the capital A art world is going. And they long ago gave up abstraction. Depending on how involved you are, most people think that modern art is abstraction, but there really hasn't been anything along the line of abstraction on the cutting edge since the 1960s. In the '70s and '80s there's been a move towards realism but it has to be unpleasant, distasteful, hard to swallow. In my twenties I was a cubist, then an abstract expressionist. That's what Harold Town was doing and the Bobaks, and whoever there might be, and I was au courant. And I'm still a bit in my own way being au courant even though the art snobs and art establishment don't recognize me as part of their family, AT ALL. However, in my view I'm being au courant with what's being done in the world of art. And doing distasteful things that nobody would ever want to hang on their wall.'

Except that because they are Batemans, and because they are beautiful of themselves, even though the content is distasteful – the clear-cut forest, the whale caught in a net – people do want to hang them on their walls. This is why the Bobby Bateman who grew into Robert Bateman is less free to follow the contours on a topographical map to a lake he hasn't seen before and is instead invited to accompany or lead expeditions to exotic places on the globe, where he meets with people concerned with the decline of panda populations, or, in Indonesia, the pressures on the Sarawak people.

Still he does manage to fit in what he calls minute exploratory trips near his home on Salt Spring Island, where, when he's home, he takes a hike each noon or in the evening. 'I've got certain trails I've made. They're not secret trails but they're my own little trails, with little minor beauty spots along the way, and one might be an open sunny glade and another might be where a great big old tree was cut down and left and it's all gone to moss, and you can see the stump. Another is a cathedral-like setting with a HUGE broadleaf maple and alders all around the outside and it's all empty in space like the inside of a cathedral and then a total carpet of sword fern underneath so in the summer you have this lovely leafy green cathedral arch and dome. And I almost tend to say a little prayer. Every time I go there I stop and I'm quiet and I look at it and I do my breathing a little bit, and I try to take it all in, the essence of it, or places along the shore. So I do that almost on a daily basis at that level.'

Hints of how he perceives his surroundings may be gleaned from the instructions he gave to students accompanying him on a trip, described in *An Artist in Nature*: 'During a walk up a mountain stream bed a few hundred yards from our campsite, with the sunlight glinting through the giant spruce trees, he urged us to look downwards at the rocks, lichen, mosses and ferns, to appreciate the different effects of light sources. "Turn in a full circle looking downwards a few yards ahead of you. You can see that when the light is coming from behind you – the usual direction for amateur photographers – all this thick vegetation looks like so much coleslaw. But if you turn so that a photograph would be back-lit or side-lit, the ingredients become isolated and big forms and patterns emerge. Now an interesting picture becomes possible."'

Our rambling conversation returns to his concern about the lack of regard in most countries for the destruction of wildlife habitat in Asia and Africa, the trade in animal parts, and wasteful fishing methods, all of which back up his statement that how people perceive wildlife and their relationship to it is the most important question facing the planet. This is Bateman the naturalist, the conservationist, the scientist-manqué speaking. Portraying wildlife, especially in the way he does, eye to eye, with respect, should lead to a greater awareness and concern for the creatures with which we share the earth, I suggest. But he claims this is not why he paints.

'No. To be honest and truthful,' he laughs yet again, 'no. If you were a newspaper reporter I might, just to make them happy, say yeah, but I really do it totally for self-indulgence. I just love to do it. I mean I'm getting a charge out of doing this wolf disappearing. And I've been offered already to go to Northern Mexico and stay in a wealthy aristocrat's hacienda right in the middle of the Sierra Madre, where these wolves occur, and this wildlife biologist will go down there, it's a woman, and she will take me out, and we can live as guests at this guy's hacienda, and you know it would be great, but I can't get it into my calendar. And I don't want to say it's selfishness, but my primary motive is I do it because I love it and I can't stop doing it and I'm all excited by it.

'And then the other, if I can make a bit of money or raise consciousness or something by my art and my prints, then I want to do that because I don't want to say noblesse oblige, but I've been given so much and if I can be of help ... It goes back to my mom. I've got this strong boy scout United Church woman in me that wants to do good. So that's also motivating. But I think it's minor what us wildlife artists are doing to change consciousness.

'In all honesty, I think TV is way way more important in a good way. TV is also way more important in a bad way, but in *our* culture, once again, a lot of the people in their forties and fifties grew up watching Marlin Perkins and *The Wonderful World of Disney*, and every Sunday evening you used to sit on the rug watching TV, so they're imprinted, like Konrad Lorenz's geese, on the fact that nature is wonderful and varied and fun and good, with a capital G, and also endangered and threatened. And with all the stupid mistakes and garbage and cheating that both of them did, it doesn't matter.

'I didn't sell any paintings until I was thirty-five. I just did it because I couldn't stop, and it wasn't easy either. I remember a fellow teacher. I didn't have a car when I started teaching, in fact I was quite slow socially. I lived at home with my parents 'til I got married at the age of thirty. It was a free house and I couldn't see any reason to get away, I had them quite well trained,' he laughs at the arrogance of his statement. 'And so I'd get a ride home to Toronto and this teacher said, do I envy you, going home for the weekend and having a lovely relaxing weekend with your hobby of painting. And I said to him, painting ain't a hobby. I teach for fun and I paint for real. I just happen to be paid for the teaching. I said that's like saying to Sam Snead or Jack Nicklaus, did you have a relaxing weekend at the Hollywood Open? It's never been relaxing for me. It's driven me crazy and into a state of despair and depression. But it's what I do. It's what I can't stop doing, and it's just always been that way and I can't explain it and I can't justify it.'

SUMMER TURNS TO FALL, and the snow geese return. It's early November and I'm drinking coffee with Robert Husband, in the kitchen of the house he shares with his wife Betty on the Westham Island farm Betty's grandfather established. Robert, a big man of seventy, whose hairline begins low on his forehead and whose bulk and gentle energy make him seem much younger, is pointing out some of the artwork on the walls of the living room and the kitchen. He has a number of paintings, including several of snow geese, that were painted by the late Hugh Monahan, who was at one time the wild bird artist for the Natural History Museum in Ottawa.

'These are Monahan's. There are two in there of snow geese that he painted out here. Well he's gone now. Oh yeah. Died out here. I found him sittin' down like this with his shot gun in his arms and it pourin' rain. Well the neighbour phoned me and he said there's something wrong, that Hugh hasn't come in. I guess he was a little scared to go out, so I went out there and that's the way I found him. It was just about straight out here,' he says, pointing west. 'The neighbour's farm. He lived in North Vancouver and came out to hunt and paint. That day he phoned me from the Chinese store down there, Chung Chuck's. He come around the corner and they don't know if he blacked out or what but there was a big log there and he straddled it with his car. So they phoned for me to go down and get him, cause I guess he was pretty shook up and he give them my name. But we were out so he gave them the neighbour's name and they went down and got him and he came in here and he seemed alright, didn't he?'

Betty calls her agreement from the living room.

'He always came in for a bite or something. And then he went hunting. And uh ... as I say Johnny phoned, and I went out in the field and there he was, his dog – a yellow lab – beside him, just like that. The doctor said he never felt nothing. He just felt tired. It was perfect for him because the snow geese were flyin', the ducks were flyin'. And the

Robert Husband with Hugh Monahan painting

pourin' rain. You know he liked that, being out there in that kind of weather. He liked hunting, he liked painting. Well then we had to come back and report it, and we waited for the police to come. We went back with a wagon and a tractor. The policeman said we can't touch him until the coroner comes. And we waited and we waited and finally he says to me, you know, he says, this is stupid. He says I'm here to witness anyways, so he said, do you want to take the shotgun out of his arms and lay him down? Well that's what I said, let's lay him down and cover him up. Even though he probably don't feel it, but ... he *doesn't* feel it. So anyway I took the gun and unloaded it and set it on the wagon. And we waited for a little bit longer and nobody showed up. Nobody showed up so we picked him up and put him on the wagon and took him to the yard. They were waiting for us there.'

# 7

# Down the Ditch!
# Dialogues with Hunters

HENRY: I almost drowned once. I fell overboard, chasing a crippled goose. The geese were swimming with the wind – they go pretty fast – and I started poling after and poling after.

GIL: See, when you're in a punt you're standing up and poling. So if that pole sticks in the bottom and the punt's going, you don't know whether to ...

HENRY: The way it hit me, kitty corner, it just kind of knocked me off balance and I went over backwards. But I had my pole tied to my hand, eh? Virtually that's what saved my life.

PAT: Lots of people have drowned out here. Hunters. People that we don't necessarily know. I met Henry at the barn over there, or halfway down the field, just after he'd gone through that experience. There was water all around him. He just managed to get his end of the punt pole. Wasn't that how it worked?

HENRY: My chest waders filled with water and I used the pole to push myself up for air. I was probably in the water twenty minutes, but that was too long for me. I had drifted from our slough, where we hunt, to the long point where the snow geese land. That's where I was.

GIL: Not Pelly Point.

HENRY: Yep.

GIL: Way out there?

HENRY: Yeah. I was drifting, because I couldn't get out. Hey, I talked to that guy up there, I'll tell you. I almost left the punt. They have those big snow goose scotchmen out there? Those big green balls that mark the place where they fenced off sections of the marsh so the snow geese couldn't get at them? The fishermen call them scotchmen. I saw the last one. I looked at it. I looked at the punt. There was no way I was going to let go. I think my fingerprints are still in the fibreglass, I'll tell ya. And I thought, if I can touch bottom I'll get to that scotchman. I didn't realize how bad the hypothermia was in my legs.

It's a cold, bright November day, the 17th. There was ice on the Westham Island bridge this morning, and I am somewhere I have never been before, alongside a ditch in a farmer's field, listening to hunters' stories. The party I'm with includes Pat Mulligan, a former conservation officer and father of a well-known Canadian television personality, Terry David Mulligan. Pat is eighty now and perambulates with the assistance of crutches or a walking stick. Today he leans on the root end of a cane of California bamboo, and when he has to move from the hummock he has been resting on, the two younger men, Gilbert Saunier, a retired phone company employee in his early sixties, and fifty-three-year-old Henry Parker, a designer of cranberry-harvesting equipment, are at his side in seconds to help. The fourth member of the group is Cory May, Henry's fourteen-year-old neighbour whose dad owns and operates a large cranberry farm in Richmond and sits on the board of Ocean Spray.

Today is the first time I've met Cory and Gilbert, but it is my second encounter with Henry and Pat. I was talking with Robert and Betty Husband in the kitchen of the house where they live on their Westham

Island farm a couple of weeks ago when Henry came in with a big sack of fresh cranberries for Betty. Robert told him I was interested in snow geese, and Henry contributed a few observations, and when I drove out, shortly after he left, I found him idling at the crossroads. An excitable, ruddy-cheeked, black-haired, black-bearded man, he stuck his head out the window of his truck and asked: 'Do you want to see some snow goose decoys, before they get all messed up?'

I drove behind him down the road to a barn on the edge of a corn-field, climbed out of my car, and followed him inside. There on the floor lay seventy freshly painted fibreglass snow geese, inert in their various positions, spots of black near the rump to simulate the folded-back pri-maries, coral-red bills and feet. I think it says something about how snow geese perceive the world that these decoys can lure them into a field, or down onto a marsh where hunters wait in their punts.

But this is what the three men and the boy are hoping for today. Henry set these seventy, plus about fifty others, out in this field yesterday and covered them overnight, so this morning they are still moist with dew that glistens in the sunlight. If the snow geese fly over and spy a raft of shining white goose shapes they will, Henry claims, say, 'Goodbye, guys.' But as we stand and talk through the morning and the sun burns more strongly down through the atmosphere, the dew gradually evaporates. Henry, Pat, Gil, and Cory have been here since 7:30. I arrived an hour later, and by ten we still have not seen a single snow goose. Pat has a doc-tor's appointment at two today and another tomorrow at nine.

'Well then, let's hope they fly before two today and after nine tomor-row,' says Gil, adding, for my benefit: 'Once they fly it's often over quite quickly.'

He's a courteous man, consistently conscious of the fact that I'm a greenhorn here: he explains, or asks Henry to explain, terms with which I am not familiar, such as corking. Pat has been telling a story about waterfowl hunting on a lake in the interior of the province many decades ago. 'Does Mary know what corking is?' Gil asks.

'That means when someone comes in and places their decoys in front of you,' says Henry. 'It's a no-no.'

Henry seems more subdued today than the day I met him, when his

Pat Mulligan hunting

enthusiasm for someone interested in snow geese launched several stories, including an abbreviated version of the one that opens this chapter. We were standing at the side of the barn talking that day when Pat arrived. Both Robert Husband and Henry had urged me to look up Pat, a long-time hunter and game warden who has tales that go back to the days when market hunters still worked the Fraser Delta. He swung out of his car, and Henry introduced us, and as we stood talking Pat donned layers of clothes, pulled a wool toque over his white hair, and finally grabbed his crutches from the back seat of his car. It amazed me that a man in his physical condition would still want to exert himself in this way. But his spirit appeared to be as strong as ever. When I mentioned that Robert had demonstrated his goose call for me, Henry agreed to try his. All these men are mouth callers and proud of it, disdainful of those who rely on mechanical callers to bring in the birds. 'You get Henry and Robert together, two different octaves almost, and they really come down,' said Pat, obviously proud of his friends.

Henry had his dog Tar with him, a female black Labrador retriever, and to demonstrate her retrieving ability, Henry instructed her to 'go get the duck.' She leapt into the back of his truck and returned in seconds with a wigeon he had shot the night before. Henry praised her and replaced the duck and brought out a couple of ziplock bags, snow goose jerky in one, craisins – dried cranberries – in the other. He offered me some, Pat some.

'Better than Thanksgiving dinner, eh?'

I knew my snow goose story would not be complete without the views of hunters, which is why I went to see Robert Husband. But the hour or so we conversed in his kitchen, drinking Betty's well-brewed coffee and munching her homemade cookies, provided mostly a retrospective view. Now I was right into it, the dog, the dead duck, and a chance to try the meat of a snow goose. I feigned casual appreciation, but this was a big moment for me. I thought of my first communion. Eating the flesh of a creature which had, until now, fed me only in the spiritual sense.

A few minutes later I thanked Henry and Pat and made a date to go hunting with them in a couple of weeks, then drove down the road to the sanctuary, where I spent the next hour or so watching snow geese in the

Thanksgiving field as I digested the conversations I'd had that morning and early afternoon. Seventy-year-old Robert Husband and his oldest son Kevin raise turnips and cabbage, strawberries and cattle on land that has been in Betty's family since before the turn of the century. Robert has lived on Westham since he married Betty and has watched snow geese come and go for over fifty years. Betty remembered having watched them arrive each fall since she was a child. Robert had raised snow geese and other geese and ducks in his backyard, punted around the sloughs after hunting season collecting cripples. He served on the board of the Reifel Sanctuary for over thirty-one years; kept records for the Wildlife Service for thirty years, recording especially the day the snows left to migrate north; banded swallows and observed their return. And he is a dedicated hunter. 'Devout,' is the word Pat Mulligan used, as if he were describing the member of a religion.

Though twenty years younger, Henry Parker has had a similarly intense involvement with snow geese. He grew up in Richmond in the '40s and '50s before that part of the delta was paved over with highways, shopping malls, housing developments, and he remembers playing hooky with friends who liked to cycle out to the dykes to hunt and trap. He raised wild birds in his backyard for a number of years and watched his pair of snow geese rear broods every spring. He too served on the board of the Reifel Sanctuary, he and Robert donate food for the annual pig and corn roast put on by the BC Waterfowl Society, and he's contributed to the operation of the sanctuary in an even more prominent way. A structural iron worker before he injured his legs, he built the observation tower people climb to look over the sloughs and ponds, out to the marshes beyond the dyke, and the foreshore of Georgia Strait. The same tower I've stood on dozens of times, where I sat to interview the sanctuary manager, John Ireland, while he painted the steps a steel grey. And back when the sanctuary kept a flock of snow geese for display, Henry and Robert were responsible for bringing them in. 'We'd catch all the cripples that were out there and take them in and fix them up. They'd live for years. The flock went up to thirty in there, all birds we'd picked up.'

Though I have a brother-in-law who hunts, and I lived in the Yukon Territory for several years, where 'getting a moose' was as much a part of the seasonal norm as getting a Christmas tree, I've been a city dweller for most of my life and congregate mostly with other city dwellers. I am one of those people Bob Bateman described, who sat in front of the TV on Sunday evenings watching Walt Disney and Marlin Perkins, who grew up to believe that 'wildlife was good with a capital G.' I am trying to follow this project with an open mind, even though what I have learned about perception tells me that there is no such thing as an 'objective' view. And it's true that I have heard stories about hunting atrocities, just recently a farmer on Brunswick Point who went into his field and blasted ten snow geese to death. I recall the hunters I sat among at the Scottish Cultural Centre, who called the scientific research the government was handing out more of the same old crap.

So I was unprepared for the impression that was rising so strongly in my mind after talking to Robert, Henry, Pat. Their involvement is passionate, of the blood. They have raised and often rescued snow geese; shot, eaten, and almost been killed in their pursuit of them. In a way that is not true of scientists – at least not the scientists I talked to – and may or may not be true of artists, they are interacting with snow geese creature to creature. The lyrical white birds with the ebony wing tips that were feeding and chattering, flying up and down in tranquil safety in the sanctuary field across from where I sat thinking, were and would continue to be engaged in almost daily battle with men like Robert, Henry, and old Pat Mulligan.

Hunting season opened on 8 October, before the main flock arrived, and will continue in British Columbia until 27 November. It is now November 17th, and Henry is the only one of the hunters who has shot any geese at all this season, five of them, the daily bag limit, yesterday. It is not because there are no birds around. Sean Boyd's November 4th census revealed that almost 50,000 birds were in the area, more of them on the Fraser Delta than the Skagit. Pat wonders if they fly down to the Skagit because 'they're getting the hell shot out of them here,' but Henry doesn't think so, and Sean Boyd's surveys prove him right. A higher percentage of birds have been staying longer in the Fraser each year than they used to, presumably because they have an upland refuge, a place to escape hunters when high tide covers the bulrush zone. Even so, we've been standing here for a couple of hours and we haven't seen a snow goose yet. No one seems worried about it. The sky is forget-me-not blue, and one of Tar's puppies, a six-month-old chocolate Lab named Mocha, amuses us by sniffing after real or imagined creatures in the grass. Pat passes around a bag of candy, we all take out our thermoses. Cory, dressed in a black sweatshirt printed with a Hard Rock Cafe logo, seems content to listen to the stories the older men tell. There are not many opportunities for boys Cory's age to get out like this, Henry tells me later, for not many hunters have snow goose 'outfits' any more, the string of decoys, the experience. What he shoots today Cory will clean and cook himself, for his mother won't touch a game bird. Nor are his brothers interested in hunting. They prefer to spend their time in shopping malls, says Cory. 'What a pathetic way to spend your life!'

In the lazy manner the morning has assumed, the talk drifts to how modern agricultural methods have changed waterfowl habitat everywhere, but particularly in California where water shortages encourage innovation. 'There used to be really great marshes,' says Pat. 'But everything's dead. Even the big trees are dead.'

This is because of a practice called laser levelling, explained by Henry – at Gil's reminder – as a way of levelling a farm field with a laser attached to a D-7 or D-8 cat that takes off the humps and fills in the holes so that land is left perfectly flat. 'You take 25,000 acres that's laser levelled, you don't need very much water to grow your rice.'

Some of the land around here has been laser levelled too. Henry points to a few fields within our vision, and as I look across the grass to where the decoys sit I see that the dew has dried on most of them. If the snow geese fly over now, they shouldn't be put off by any shiny bird shapes. But it's after eleven and Pat is getting stiff. 'I got to try to stand up for a little bit,' he says. Henry and Gil haul him onto his feet, and he leans on the root of his California bamboo cane as he recalls another snow goose story: 'I want to tell you about a period of time many years ago down here. The foreshore was quite a bit different. There was no Reifel Sanctuary or anything. There was a good many thousand snow geese down here. And we got a cold spell. Now what would happen, the tide would come in, and it would be freezing cold all day long, and it would freeze the marsh, all the bulrushes, everything was covered with ice about that much. Well the tide would go out, but the ice would stay, and I actually saw the geese land on the outside of the ice, and they would walk in underneath the ice and be feeding under the ice. Yeah. I actually saw it. They were so hungry. And the geese used to be quite poor in those days. What I mean is, later on, in the latter part of the season, if you shot a goose, nine times out of ten you had a ridge bone indicating that he was hungry.'

'Nobody would shoot them,' says Henry.

'They were no good as a table bird,' Gil explains.

I am tempted to ask Pat why he persists in coming out when his health is so bad, when it takes such an effort just to get back and forth to his car, and as though he's reading my mind he says: 'You might wonder why a person getting on in years would still want to hunt snow geese and hunt period and why they'd want their ashes spread out here. Well if I were going to die, I'd like to do it out here. You know the guys'd have to shoot and drag Paddy, shoot and drag Paddy. It's only a one-way trip.'

Gil and Henry laugh, and Henry talks about some of the old hunters who have asked that their ashes be spread in the marsh. 'You put their ashes in the shotgun shell and fire them out there. Russell Young is out there,' he says. 'John Dixon.'

PAT: And Mulligan. They're going to save enough out of my ashes, and somebody will maybe fire a shot out there.

GIL: A lot of people spent a lot of hours out there.

PAT: Some of my best times have been out here.

CORY: I feel free when I'm out in the wild.

So what's the attraction I want to know. Is it the pursuit? The literal hunt?

HENRY: Sure. In years, say, when there are no juveniles and they're all white birds, it's a real good challenge, because they're a little smarter. But juveniles will actually drag the adults into a set of decoys many times over. Decoy shooting snow geese is a lot of fun. You're layin' out in the water out there, that's a lot of fun too. In a duck punt, flat on your back. Sometimes you stay out a long time.

PAT: There's something you should try, Mary. Get Gilbert to take you out in a punt. That's all they ever used to do when I got started.

You're lying flat on your back – that's why everyone has arthritis. Rain's beatin' down on your face, you got a little bit of breeze, your boat's rockin'. You gotta be able to lift your feet up and swing around if they start coming from the opposite direction.

Henry chuckles as if acknowledging that some people wouldn't consider an uncomfortable experience such as that 'fun.'

Gil adds, 'That's the way I really started to hunt, Pat, and I still prefer that method. Maybe it's not as productive. Put it this way: for the number of birds you get you put in far more effort.'

But they seem to like the effort. Many of the hunting stories I've heard feature this theme, the difficulty, the bad weather, the survival challenge, and I wonder if hunting – the desire to hunt – stems from instincts honed in our evolutionary past: the heightening of the senses, the accomplishment of survival, that hard clean feeling of having made it through a tough experience.

The missionary H.R. Thornton described two different attitudes towards hunting in the late 1800s in northern Alaska: 'Oddly enough, the Eskimo seems to feel none of the true sportsman's enthusiasm: there is no wild delicious thrill, such as comes to the Anglo-Saxon gunner – at least when he sees his game before him: when everything else in the universe is blotted out, as it were, for the time being, and he is aware only of himself and of the swift bird or the cunning (and perhaps ferocious) beast before him; when every muscle is tense, every nerve tingling with energy and excitement, every sense sharpened and quickened a hundredfold; when he is all ear to hear the slightest movement on the part of his game, all eye to detect its position or anticipate the direction of its flight, all hand to wield his weapon with lightning-like rapidity and the utmost precision; when all thought of the suffering he may cause is swallowed up in the fierce, glad conflict between man's skill and courage on the one hand and the animal's swiftness, cunning or ferocity on the other. As we have said, the native hunter appears to experience none of these sensations. Whether this is due to his phlegmatic temperament or to his ardor being dulled by constant habit, seeing that hunting constitutes his daily work, or simply to the fact that he is less demonstrative than his Caucasian brother, we cannot say with certainty.'

The differences he noted between hunting for necessity and hunting for sport is something that was recognized as far back as the Middle Ages when it was assumed that those who hunted for meat would use the most efficient method possible – netting fowl, for example, rather than shooting it. Those who hunted for their own pleasure were expected to adhere to the code of the 'sportsman,' which included giving prey a fair chance to escape and ensuring that wild creatures did not suffer unnecessarily by tracking down and killing those animals that had been wounded.

What began as necessity continued as challenge, thrill, and something more: the socializing, the pleasure of being outside in beautiful natural surroundings. 'Male bonding, isn't it?' Gil laughs a little self-consciously at the new-age phrase, but it's true that none of the wives accompany them on their hunts, though Henry's grown-up daughters used to like to go out with him when they were younger.

'I'll tell you,' says Pat, 'the guys in this hunting fraternity, if they say something, it's as good as a handshake or a contract. There's no b.s.-ing. They're right down-to-earth types that know how to get along in this world. I trust them as people and as friends. And that's a hell of a lot more than you can say for most people. We do on occasion lay our life on the line for one another. We've all had close escapes out on those punts.'

The sun is nearing its zenith, and Henry's dog, Mocha, is bored, whimpering. The talk turns from stories of survival to guns, especially the one Pat uses, which is, he claims, the oldest Browning shotgun being used in North America. The date of the patent is 17 December 1901 and the serial number is 174. It's an automatic shotgun, unlike Henry's and Gil's which are pump action, a Remington 870 and a model 12 Winchester respectively. Henry's has a black plastic stock he made himself, to protect it from the water that gets into the bottom of his punt.

Henry prefers the pump action because it lets him mix up shells. 'You can have a low brass shell for your first shot, especially in a duck punt where the birds are really close.' Gil finds that with his pump action he can shoot three shells just as quickly as the automatic, because they're aimed shots. 'Not just bing, bing, bing, and it's not as prone to jamming.'

Pat searches for his gloves. It looks like Robert will not show up – I learn later that he wasn't feeling well today – and since he isn't here with his truck, Pat will have to walk to his car. Henry offers to carry his things for him back to the barn where he parked. But Pat says, 'I'll have another cup of coffee and then I'll go. I'm not going yet, I'm just kind of getting organized.'

Before he leaves I want to get the question that has been on my mind out into the open, so I ask him what he thinks about people who are opposed to hunting. 'They're living in a world of ignorance, I find. They don't know what they're talking about and they don't know hunters and they don't plan on knowing hunters and it's so easy to criticize. That's my feeling. I have little time for them.'

Gil adds, 'We can kind of appreciate their position but they cannot appreciate our position. We get just as mad at the irresponsible hunters as they do because they give us a bad name.'

Pat continues: 'This is something I've been doin' since I was about six years old, you know. Huntin' rabbits back on the prairies, and Mr. Oliphant, the hardware man said, Mrs. Mulligan, you're going to be sorry buyin' that boy 22 shells. A box of 22 shells were twenty-five cents in those days. And if you wanted to be really frugal you bought BB caps. My mother, being a very intelligent lady said, "Look Mr. Oliphant, if I buy them and give him the shells, I know that he's got them and he's not going to do any harm with them." And I kind of went through life like that. That if I pass through this world and didn't do any harm and did a little good, we've got a better world.'

I remind him that there are people, animal rights activists, who would claim that shooting snow geese is hardly doing the world any good, and his impatience comes out in the tone of his voice. 'Aw, they obviously haven't been to a slaughterhouse. If they want to get into an argument with hunters, they don't have a leg to stand on, because you say, lady, do you eat pork? And she says, oh yes, we eat pork. Ever been to a slaughterhouse where they kill pigs, lady? Well you should go there, lady. Or go there the day the rabbi is killing beef for kosher beef. They don't even knock them. Or chickens. They fill a room full of pigs, a couple of guys go in there with mallets about so big. And they just walk up and slug the pig on the head, and down they go. Bang. And that last pig is just bloody well climbing the walls, trying to get away. And then they slit their throats. Do you think those horses they kill for dogfood don't know what's happening? Or take all those beautiful milk cows.'

Henry feels people are opposed to hunting because most hunters don't actually need the meat. 'And that's where they kind of say, how come you guys go out and shoot all those beautiful geese? Can't you go to the store and buy a turkey? Well that's not the point. I mean if we don't get a goose today, I don't really care. I might get one tomorrow. I know we're going to get a shoot going between now and Sunday.'

I think of Robert Bateman's remarks on the killing of individual animals. Because he has a bit of a name, as he puts it, he's always getting requests from animal rights people for support. 'I kind of say more power to them but that's not me. I do care if an individual animal suffers, just because I'm somewhat tender-hearted, but I wouldn't put a lot of effort to it. But an entire ecosystem, now that's a different matter. If one wants to pick villains in this world that are hurting animals, trappers and hunters would not be at the top of the list. Farmers would, however, and also industrialists, and I don't see animal rights people attacking farmers. Draining marshes and swamps and killing all those baby ducklings, and lowering water tables through irrigation and spraying pesticides. The death and agony of certain insects. If you feel sorry for a baby seal, should you feel sorry for a baby predatory beetle? '

While we have no right, he feels, to tell another to kill or not to kill individual animals, he feels we do have a moral obligation to make sure that we don't wipe out an entire species.

Fred Cooke expressed a similar sentiment. Although he has been a birder since he was a boy, when he speaks of knowing an individual snow goose he means he knows its life history, especially its reproductive history, from data collected in the field. Neither Bateman nor Cooke are hunters, but both have made consumptive use, as it is called, of wildlife for study purposes. Bateman trapped small mammals for museums, and

Fred Cooke's team shot geese at La Pérouse Bay to determine, in one study, how the mate would react if her gander was killed just as she was incubating their eggs. This is something Barbara Ganter said would never happen in Europe, and when she came here she was very much against hunters. 'European biologists won't have anything to do with hunters. I hated them at home. There's so little room. There's no land really and there's so many people and so few animals and to then start shooting at them. It's really disturbing. At least on the coast, on water birds, they can't be hunted any more where I worked. And so biologists just can't stand hunters in Europe.'

But European hunters come from a different cultural tradition. In feudal times the right to hunt belonged to landowners, the aristocracy, including royalty who gave each other titles like 'Lord High Master of the Chase.' The aristocratic tradition continues through shooting clubs, which, though their memberships may be relatively large, still retain a flavour of exclusivity. One thing that makes North America different is that here any one who can buy a licence can hunt.

'At first I had absolutely no understanding, no acceptance of hunting in North America, and then I slowly started seeing that things are different. There is a lot of room, there is a lot of wildlife and it doesn't really matter that much if some of them get shot,' Barbara said.

She also acknowledged that 'the whole goose research world in North America is very much funded by hunters. They're interested in knowing how many geese are up there and how many they can shoot and anything that influences the population size. And one of these things is disease and cholera, so there's a lot of hunter money going into these things, through many channels, but the interest in these birds is because of the hunters.'

It is a concept that may be better understood on the Skagit, in Washington State, partly because revenues from hunting and fishing licences have long gone directly into managing wildlife and acquiring and maintaining wildlife habitat. According to Mike Davison, Washington has a higher percentage of protected lands than most other states in the union. In British Columbia, licence fees are swallowed by the general revenue fund, though there is now a surcharge of five dollars applied, money that goes into a fund marked for habitat acquisition and protection. Both hunters and scientists in the province perceive a greater impact by hunters on the Skagit. The two Barbaras made a trip down to Skagit Bay in the fall and claimed they couldn't see the geese for the hunters. There may be more people hunting, but because the bag limit is lower in Washington, three as opposed to five per day, similar numbers of birds are killed. A comparison of figures I obtained from the Washington State Fish and Wildlife Department and the Canadian Wildlife Service showed that in 1994, while 1,078 snow geese were shot in Washington, most on the Skagit, 1,076 were shot in British Columbia in the fall, and another 546 in the spring of 1995. Twenty years ago there was a bigger difference. Two thousand shot on the Skagit, versus 44 on the Fraser. But in 1990, hunters in British Columbia actually shot more than twice as many: 573 on the Fraser, versus 250 on the Skagit. Numbers rise and fall for a number of reasons, but the success of the hatch is a big one. In 1993, when juveniles made up such a large proportion – 40 per cent – of the flocks travelling south, nearly 1,859 were shot in Washington, 2,233 in British Columbia, most of them inexperienced juveniles.

The number of harvested birds changes from year to year all up and down the flyway. But hunting in general has declined over the years: changing sentiments and tighter regulations have reduced the amount of hunting in the Skagit by close to 40 per cent, in Mike Davison's estimation. In British Columbia, 5,300 licences were issued to hunters in the Fraser Valley in 1976, compared to only 3,100 in 1993-4.

Graham Cooch, who watched the fluctuation in waterfowl populations for years from his various positions in the Canadian Wildlife Service, and who determined how much hunting in Canada various waterfowl populations could withstand, says, 'The number of sport hunters has declined to a level I can't believe, from 450,000 licensed hunters in Canada to 250,000 in the last ten years. The same thing is happening in the United States. Places to hunt are too hard to find. There's an anti-blood-sport emphasis coming around as well.'

Cooch and others of his generation and those after him saw hunting as a 'management tool.' Where humans are not involved, snow goose populations self-regulate. When the population grows too large for the available habitat, birds starve to death. In years when the weather is bad, the whole hatch can be lost. Similarly, when lemming populations are low on Wrangel Island, arctic foxes, which subsist largely on lemmings, take more snow goose eggs and goslings instead. These dynamics existed before humans came into the picture. Aboriginal people complicated the relationship between snow geese, their terrestrial and avian predators, the weather and the vegetation by hunting them in spring and fall and gathering their eggs. Pioneering Europeans in North America increased the hunting pressure over the years, as they too subsisted on the once abundant game. But hunting did not affect waterfowl populations as seriously as habitat destruction. Wetlands disappeared, leaving fewer places for geese to feed in the winter, meaning there were some 'extra,' which could be shot by sport hunters. If they weren't shot, they would die anyway because their numbers were outstripping the ability of the available land to support them. So wildlife 'managers' came into the picture, to survey numbers and reproductive trends and decide how many geese hunters could shoot without threatening the viability of the species.

The concept of managing, controlling something that was doing fine before humans entered the picture strikes some people as offensively human-centred; others consider it the only responsible way to coexist with our fellow creatures on the planet. But the return-to-nature movement, the antihunting sentiment is having consequences that Graham Cooch considers dangerous. He refers to the problems of snow goose overpopulation that have developed in the eastern Canadian Arctic.

'In many areas Eskimos have moved in off the land and are living in settlements. Queen Maud Gulf has no Eskimo population at all, and so the harvest is negligible. Fifteen hundred birds shot in New Mexico out of a wintering population of 75,000. Maybe 3,000-4,000 shot out of 125,000 in Chihuahua. We are sitting on a time bomb of fowl cholera and habitat destruction. It could collapse in my lifetime. Certainly in the lifetime of people like Evan and Fred Cooke.' And as Fred Cooke found, it is not only the decrease in hunting that has allowed the populations to increase, but also the fact that farmers leave more waste crops on the ground, which waterfowl feed on in winter.

Because weather is such a crucial population determinant in the case of the Wrangel nesters, the Wrangel group is holding its own, even declining a bit. So overpopulation is not a problem on the Fraser and Skagit. It's a different story, though, for the portion of Wrangel nesters that winters in California with birds that nest on Banks Island, in the Canadian Arctic. There have been outbreaks of avian cholera in California, quite possibly because the disease spreads quickly in more crowded conditions, just as cold viruses spread in schoolrooms. Wetlands in California decreased 90 to 95 per cent from the late 1800s through to 1980. But since then, Greg Mensik of the Sacramento National Wildlife Refuge has noticed more stability and even a slight increase in habitat for waterfowl as rice farmers, instead of burning off rice stubble, began flooding it off. So crowding may not worsen, but whether it is agricultural practices or some other element in the environment that promotes the spread of avian cholera has yet to be determined.

It's NOON, and Pat has to leave. Henry slings the older man's gear onto his back and helps him cross the field, the equivalent of, perhaps, three city blocks, to the barn. The dog follows them.

We still have seen no geese and I'm wondering if there will be a hunt today, but soon after Henry returns from seeing Pat to his car the wind changes, there's a taste of dampness, of salt to it and Gil and Henry hear geese up over the marsh, perhaps a mile in the distance. Henry has left young Mocha in his truck and returned with Tar, Mocha's mother, the more experienced hunting dog. Tar sniffs around the ditch and comes back with a weak, injured mallard that has been lingering in the ditch, apparently unable to fly out. The blade sharpness of its keel shows it is starving to death. Henry walks away from us and quickly dispatches it, the polite term for wringing the neck of a bird too far gone to survive.

It is close to one o'clock when we finally see a group of five snow geese sailing over from the sanctuary. Henry slips his camouflage jacket on and tells the dog to get down. I crouch in the grass, next to the hummock that Pat occupied for a few hours this morning. It's an odd feeling, as if I'm a kid trying to trick somebody, or a fugitive being pursued. Henry falls onto his back and fires as the geese fly overhead, and the dog leaps up, but, no luck. Henry's gun jammed and the dog ran out into the field.

Even so, neither the sound of the shot nor the presence of the dog seems to have frightened off the geese.

But these have been only the advance group, scouting it seems, for minutes later the hunters spot more geese coming our way. 'We better get serious about what we're doing,' says Gil, taking his position. They begin to call again, 'Ah, coo, aah, oo!' But the group stays far to the north, out of range.

As if attention is a pencil sharpened by use, the focus of the group dulls when there are no geese to track. Gil notices swallows, a robin. He mentions a goose he saw whose black primaries were especially prominent. Henry rehashes his moves. 'I should have let them come over again,' he says. Gil philosophizes: 'It's a judgment call, always a judgment call.'

But the geese are definitely more active, we can hear them rising off the marsh, so loud that Cory thinks they sound like a big train. Henry tries calling again, 'Wah, coo whah, coo, wah coo.' Gil tries. A group of fifteen to twenty flies over high, I tense, as if I can feel the shot about to explode. But the birds just keep gliding east over the field and curl back way out of range.

'They got sharp eyes lookin' down at us. The bigger the bunch, the sharper the eyes. There's always an experienced one in there,' Gil muses.

With eyes on either side of their head snow geese have only monocular vision: they can see objects on the sides of their bodies better than those straight ahead. They also have superior distance vision and can distinguish movement. While they will ignore waving branches and grasses, they immediately spot the movement of a living person or animal in their environment. But none of the books I read told me how a man appears to a goose. Images that fall upon the retina must be interpreted, and since birds need only recognize others of the same species, to reproduce, and threats to their safety, for all we know a man and a coyote might look the same to a snow goose.

I wonder aloud if it's hard to aim at a small group, such as the one that just passed us by, if the hunters have better luck with the big flocks.

'There was twenty, twenty-five in that group. We shoot at fours and fives, threes,' says Henry.

'See,' says Gil, 'You still have to pick out the bird. You don't flock shoot. You have to single out the bird. That's why you often shoot better if there's one bird there. You're aimin' at one, picking out the next one as it flies over. But what happened with that group, they went right over us with their wings set, eh? And if they'd acted correctly, they would have gone down about a hundred yards and come back into the wind, and then there would have been a shot and they would have probably tried to set down just on the other side of those decoys.'

'The wind's pickin' up, you know,' Henry observes.

And the mood is changing too. Gil suggests that we move into the tree shadows and hunker down. At the sight of a skein moving in over a barn to the west, we do. This is a big group. A hundred? More? Gil and Henry, and even Cory I think, renew their calls, all trying to imitate the houck, houck, ou, ou sound meant to tell snow geese that there are others here, it is safe. Do the houck, houcks, the calls I have heard as songs, choruses, mean different things, as the sounds of whales, the songs of birds do, I wonder? And if so, do these hunters actually understand the meaning of the sounds, or only mimic them?

Then, 'They're comin' right at us, Gilbert.'

'Yeah I see them. There's two bunches.'

'Probably more of them comin' with this wind. See? It's really picked up.'

They start calling again, and as the geese fly near, the voices of the men are mixed with the voices of the geese. And then other sounds. The crunch of a shot. The thud of a snow goose hitting the ground, just across the ditch. Bang, thud. It's the sound that will stay with me strongest, even though more geese will be shot in the next hour. Even though it will get quite frenzied, as more and more geese crowd the sky above us, flying directly towards the decoys though they must sense, now, that this is not a safe place. I consider why they do this even as Tar splashes back across the ditch with the first snow goose I've seen shot, and drops it at the foot of her master, Henry. It is a dark grey juvenile, one of the goslings that broke out of its shell in early July. Its deeply rust-stained face shows that it quickly learned to grub on the marsh at the edge of this island where we are crouched alongside a ditch.

The sky is still blue, the sun whitely bright, the wind with the sharp, slightly damp flavour it picks up from the sea. There's a real sense of waiting now.

'My way, my side!' Gil shouts, as the men start to call again.

We've seen only a portion of the flock so far, but the big group is on its way. I can hear them over the snore of plane engines high above.

'Here they come over Andy's. Look at this. Mary, look over Andy's barn there. Straight away? See them all? Cory, see them? Hundreds of them now.'

And it looks like they're heading straight for us. Gil suggests that we nip into better positions, into postnoon patches of shadow cast by the bushes and weeds above the ditch. The men begin to call again, but the geese never come as far as the ditch before instinct, or whimsy directs them towards the sanctuary.

Then another wave breaks into the cerulean band between us and the sea.

'Uh-oh,' Henry says. 'Down the ditch, down the ditch! Get down!'

This group is flying low. I know the hunters are going to shoot this time and prepare myself for another thud. I count eight shots. Three geese fall with the first four.

'Oh,' says Henry, who moves from a crouch to a supine position, rotating his shotgun as white geese steam over us, 'That one got my back.'

But he keeps Tar still and the shot geese lie where they fell because more geese are coming, the men continue to call them in.

'Down the ditch!' Henry calls again, as they head straight for us.

'I need shells,' Cory calls.

'Stay there for a minute,' Henry cautions.

Excitement charges the men's voices: the calling sounds almost frantic now. As the geese lower in I hear, three, four, five, six more shots.

At least one goose is hit and plummets into a tree. The dog goes to fetch it, and Henry doesn't even try to stop her this time as the sky is literally thick with geese, an undulating sheet of them just overhead.

'Holy shit! There are so many,' says Cory.

I lose track of the number of shots, the number of geese that fall. The air is peppery with the smell of gunpowder now, homely with the wet of the dog, Tar's coat, as she races back with another dead goose.

The long retrieve, half a mile off shore:
Henry Parker and his dog Tar

Henry takes advantage of a pause in the action to dash across the ditch and retrieve one that's fallen far across in the field opposite. Tar brings back still others from the field behind us. Alongside the first juvenile there are now more geese, adult and juvenile.

I notice that one is still alive, and Gil immediately picks it up, turns his back to me and dispatches it, 'It's the most humane thing to do,' he explains. I think of the word dispatch, dispatcher: the person who sends things from one place to another, from life to death in this case, from animal to meat.

'That was pretty good shooting,' he concludes. 'They were not close. That's almost extreme range. So to knock down what we did was pretty good.'

I see the rose feet of the goose he just dispatched still swimming. 'Is it dead now?'

'Oh yeah, the nerves, you know. See? Its neck is totally severed. So it's dead.'

They count the geese lying limp on the grass.

CORY: One two three four five six seven, eight nine.

GIL: Make sure you didn't count twice.

CORY: One two three four five six seven eight.

GIL: And the one Henry's bringing in.

CORY: We shoulda had more.

GIL: Well we could have. You see we waited all day, and as you say, in ten minutes ...

CORY: We won't see any more today.

GIL: Oh don't count on that.

CORY: Unless that flock turns around and comes back. That was a big flock ...

While Gil explains to me how shotgun shells explode into pellets, Cory inspects the harvest, strangely reminding me of my daughter coming back from Hallowe'en trick or treat excursions, and spreading her loot on the carpet to count and admire. 'That's a nice big bird,' he says. 'That would be a nice one to get stuffed. Look at the white on that thing.'

Meantime Tar and Henry return, Henry with a goose whose head he cut off because, he explains, 'If you bleed them when they're still alive they're better than chicken.'

Now it is his turn to inspect the harvest. He spreads the wing of an adult goose and shows me the alula, which is still grey, indicating, perhaps, that the bird is not yet fully mature. Points out the effects of the iron oxide in the mud the goose has dipped its head into searching for rhizomes. As we hear more geese calling in the distance, Gil and Cory drape a green camouflage net over the dead geese, so that the live ones don't see them.

It's not quite two, and they don't have their limit of five birds apiece yet, but I have to leave to make the school bell that will ring at three o'clock, back in Vancouver. Henry carries an armful of dead geese to his truck, and shows me the pictures he promised to show me, of other hunts, of decoys he has carved. I don't know what I'm feeling, except tired, headachy. I've gathered feathers and down from the grass and stuffed them into my pocket. Now I ask Henry for a wing from one of the birds. He severs one from an adult, one from a juvenile and shows me where to sprinkle borax so that the decaying flesh doesn't stink. He shows me how to spread the wing and mount it. I thank him for everything and drive away, back across the Westham Island bridge, out to the freeway that connects the country with the city. I don't know what I'm feeling, exactly. The men I've just left would probably find it curious that I'm questioning my feelings at all. I've seen kindness today. I've seen what I might once have thought cruelty. So I don't know what I think today, what I feel, but I do know something I didn't know before about hunters, and that is a little of how *they* feel.

8

# Snow Geese for Supper

**20 November 1994.** Although the United States will not celebrate the holiday until later this week, and in Canada we traditionally celebrate it the first Monday of October, the house has a Thanksgiving feel to it. It must be the associations: standing at the stove, sautéing onions and celery for dressing; talk of an important football game, the Canadian Football League's Western Conference Championship, which the BC team is predicted to win.

Today my job is to cook two of the snow geese Henry shot then plucked and dressed for me and left in Robert Husband's cold room, where Robert keeps the wild fowl he harvests himself, and, more dramatically, the carcasses of big game – deer, elk, and moose – until he butchers them. When Annie and I drove out to pick up the geese Gil and Henry had offered me, Robert and Betty invited us in for coffee and cookies. Robert told me the big flock had landed in the field behind his house and spent most of the day there earlier in the week. I raised my eyebrows as if to ask, 'Did you shoot any of them,' and he shrugged his big shoulders: 'They got to eat somewhere,' he said. We sat for a spell, then Robert showed Annie his trophies, the heads of the game animals he has bagged, including one of a buffalo he shot as a result of winning a draw. He's about to take us to the cold room, and then he remembers who we are, where we come from. 'Maybe you don't want to see dead animals?' But Annie is not put off by the prospect. As for me, while I am happier to watch deer bound through the woods, and geese fly, I am grateful to Robert, and Henry and the rest of the hunting party for indulging my interest to this extent. For it seems that to fully appreciate the view of the hunter, and to experience the geese, I must eat one.

Robert hands me a bag with two birds in it. Henry has marked them with the initials J and A, so I can tell the difference between the adult and the juvenile now that they no longer have feathers to distinguish them. Henry warned me that they were hard to pluck without tearing the skin, and as I draw out the juvenile I see what he means. The plucked skin is dark, not the whitish-yellow of a chicken, but browner, and torn in a few places. But this is undoubtedly a masterful job of plucking compared to what I could have managed. When Gil suggested to Henry that I might enjoy trying a snow goose for supper, Henry thrust one of them towards me. I couldn't imagine cutting off its head, plucking it, reaching inside. I was grateful that he correctly interpreted the look on my face. This bird is now as impersonal as any I've picked off of the refrigerated counters at the supermarket, except that I know this one flew a few days ago, and I know why it isn't flying any longer.

All my comparisons are with chicken because as a cook that is the animal with which I am most familiar. The snow geese are longer bodied, their legs are skinnier. When I saw them in the sky they seemed healthy birds, and when I saw them lying on the ground I also thought them in good shape, but there isn't much fat on them, particularly the juvenile.

I wash them, and a wet smell, a grassy smell rises up from the sink. The body cavity seems larger than a chicken's, more vaulted somehow. I think of the room needed for the working muscles of a flyer. The big heart. The rib cage is composed of fine, almost delicate bones that remind me of the finely crafted ribs of a carefully built ship. No wonder I think of them sailing through the air. I stuff the cavities with a mixture of apples, onions, parsley, raisins, bread, walnuts, sage, salt and pepper, and set the oven to 250 degrees, as Betty and Robert suggested.

It was a windy morning out on Westham, and the wind has kept up all afternoon, probably explaining why the power suddenly goes out at 3:50, when the geese still have two hours left to roast. I check with my neighbour and learn that it isn't just my house, the power is out all over the neighbourhood. It will stay out for another half hour, while I try to decide whether or not to move the supper, half-roasted snow geese and all, over to David and Nancy's house, the friends I've invited to share the snow goose supper with Annie and me. Despite the power outage, the birds have been cooking long enough that the house now smells a little like Thanksgiving too. Damn!

But then, just as suddenly as it went out, the power switches on, the refrigerator motor starts humming, the goose fat and butter mixture on the bottom of the roasting pan begins to sizzle.

*Menu*
Roasted Snow Goose, stuffed
Slices of Danish Butternut Squash
Roasted Potatoes
Blackberry Conserve
Green Salad
Plum/Blueberry Crisp

Nancy returns from the swimming pool with Jimmy, six, and Hilary, three. Annie has been swimming with them while David has just witnessed the BC Lions score the touchdown that gets them into the final, the Grey Cup game they will go on to win on the strength of a last-second field goal. At 6:30 we gather at the round table in my kitchen, and join hands to give thanks for the food we are about to eat. We are not normally eaters of wildlife. We watch wildlife on television and through binoculars. Annie's and Jimmy's classes take field trips to the Reifel Sanctuary, to marvel at the beautiful white flocks of snow geese each October. We try to do what we have heard aboriginal people do, and thank the snow geese for giving their life so that we can eat. 'Snow geese?' says Jimmy sceptically.

But the girls are enthusiastic. Hilary finishes her first helping and asks for more 'chicken.' The meat is dark, tasty, and surprisingly juicy. I notice little difference between the adult and juvenile, despite what I've heard. Gil told me that given the opportunity he always aims for a juvenile, so that he doesn't break up a pair, but also because he feels the juvies are a better table bird.

David has just finished carving when the power goes off a second time that day. I run for candles. 'Now we're really living like the pioneers,' I joke with the kids.

We finish all but about a quarter of the meat, and that I boil with the bones for soup, which I share with my octogenarian neighbour, who used to hunt.

I took a picture of the carcass before it went into the soup pot, and when the bones are boiled clean I let them dry and keep them in a box on my desk so Annie's kitten can't get them. She has already gone wild over the wings Henry gave me, the feathers that share space with pencils and pens in a jar alongside my computer. My amulets are piling up. In addition to the bones, I have an old peanut butter jar filled with down, the down I collected from the field the day of the hunt, a poster advertising the Snow Goose Festival, pictures of snow geese propped here and there. An ink drawing of a flying long-necked bird on a stone. Spirit stones they're called.

# 9

# The Festival of Snow Geese

OUT AT THE Reifel Sanctuary gift shop, there are even more snow goose charms: postcards, photographs, snow goose images on tote bags, sweatshirts, fridge magnets. Lapel pins shaped like a flying snow goose. Before you even get there it is obvious that the Reifel Sanctuary and snow geese are synonymous in some way as the shape of a black-winged white goose in flight marks the signs pointing the way from Highway 17, through the town of Ladner, down River Road to the Westham Island bridge.

In the fall, the entire municipality of Delta welcomes the return of the snow geese. There are displays of decoys at the Delta Museum and Archives, a craft fair at a shopping centre, banners strung across major intersections. The municipality cooperates with the BC Waterfowl Society in arranging shuttle bus transport out to the sanctuary the weekend of the festival, to compensate for the limited parking space.

The 1994 Snow Goose Festival is the eighth annual and the fourth I've attended. In good years, and there were two in a row, visitors arriving at high tide, when the bulrush zone on the foreshore is under water, can see the snow geese flying into fields around the sanctuary where Canadian

Snow geese flying in formation

Wildlife Service and Waterfowl Society volunteers stand at telescopes ready to answer questions. The weather can be clear and cold as it was on Saturday, in 1994, or wet and grey and cold, as it was on Sunday of that year. More people come when it's sunny, of course, but the first year I joined the festival an earlier attendance record was broken, even though skies were grey and thick dampness crept through the soles of boots to chill the feet and funnelled up the spine to start shoulders shivering.

People come simply to celebrate the geese, mark their return, thrill at the spectacle of almost 50,000 white birds rising into the sky at once, singing. That the snow geese consistently draw thousands of vistors over the wintering period may be a manifestation of 'biophilia,' the word Harvard University science professor Edward O. Wilson invented to describe what he believes is our innate affinity with the natural world.

In *The Biophilia Hypothesis*, edited by Wilson and Stephen J. Kellert, Kellert presents a table that delineates the ways in which this affinity is expressed by various people. I found their book as I was nearing completion of my own investigations and was intrigued to discover that some of the questions I was stewing over have been posed formally by Wilson and Kellert and others who are searching for proof that relations with the natural world are crucial to our well being and survival as individuals and as a species.

Kellert would probably fit the hunters I talked to under his 'utilitarian' category, which describes people who make practical and material use of nature. 'Utilitarians' gain physical sustenance and security from their affiliation with the natural world. Someone like Robert Bateman would probably have largely an aesthetic affiliation, based on the physical appeal and beauty of nature, which provide humans with inspiration, harmony, peace, and security. If Bateman saw Kellert's table, however, he – and perhaps the hunters, too – might also want to weigh in under the 'naturalistic' type of affiliation, distinguished by a satisfaction from direct experience of or contact with nature, which benefits one's curiosity, outdoor skills, and mental development. And the 'humanistic' type of affiliation, characterized by strong affection, emotional attachment, 'love' for nature, which affords people the benefits of group bonding, sharing, cooperation, and companionship.

Kellert proposed nine human 'valuations' of nature, including the scientific/ecologistic, which obviously fits the men and women I introduced in Chapters 4 and 5, and the symbolic, defined as the use of nature for metaphorical expression, language, thought. This last is an easy one to appreciate if you think of how often we use phrases like 'loose as a goose,' or 'she's got a bee in her bonnet,' or 'he's like a bull in a china shop' or describe politicians as 'hawks' or 'doves.' Animals are said to be basic to the development of speech and thought, and some evidence for this comes from Kellert, who found that animals constitute more than 90 per cent of the characters employed in language acquisition and counting in children's pre-school books. Veterinarian and anthropologist Elizabeth Atwood Lawrence says, in *The Biophilia Hypothesis*, 'Animal symbolism is biophilia in that it represents another step in the age-old search for "man's place in nature."'

Many of the people I encounter on the paths of the Reifel Sanctuary during the Snow Goose Festival demonstrate that they use nature in this way, to aid in communication and mental development, every time they aim for a poetic description of the spectacle that attracted us all here. 'Look! A storm of geese!' a woman exclaims to her son.

The first day of the eighth annual snow goose festival, Saturday, islets of snow from an early surprise snowfall still cling to lawns facing north. But by noon the sun is shining in a blue, partially cloudy sky. Last year, it was warmer. A gorgeous fall day, temperature about ten to twelve Celsius, sky clear though wisps of cloud drifted in and out. High tide at 11:15, so the geese moved up to the fields around then or earlier. When we arrived they were homing in on the field to the east of the Alaksen refuge, then flying up and back, disturbed by a cruising eagle, a coyote in the fields. For two hours they kept lifting up, settling down, their loud high-pitched call like a trillion mosquitoes buzzing at once, except more exhilarating. In the sunshine, with the geese so thrillingly active, it seemed everyone was smiling. There was a real sense of celebration.

The sense of celebration is evident today, too, as soon as we enter the parking lot and are waved to a spot by one of the traffic organizers. John Ireland, in a hat with a goose head sticking out the front and wings out each side, is everywhere, helping with traffic, answering questions.

John Ireland wearing goose hat
and holding a one-winged goose

His broad Yorkshire accent would identify him even if he weren't wearing his trademark cap. More volunteers stand at the path head, offering to lead guided tours. But we say no thanks and start down the trail on our own. I'm guiding my visiting sister Mike today, a birder new to this part of the country, who has never seen snow geese before. We will find them in the Thanksgiving field, but before we reach it we take the opportunity to spot them through one of seven Bausch and Lomb scopes trained on the field. Bausch and Lomb has set up an exhibit here two years running and contributes a pair of binoculars for the draw. In the slough alongside the exhibit, I see a single snow goose, its feathers all puffed out, perched on a log with a couple of mallards.

CWS scopes, with attendants nearby, stand at several points along the path. Even more numerous are people of all ages extending palms full of wild bird seed from some of the 30,000 bags of it that are sold here each year. Chickadees decorate the brush along the path, and the bravest hop down to charm the people who have been waiting patiently with their hands out. Chickadees are the grace notes in this snow goose symphony, but mallards too look up expectantly as people pass; it's no wonder they are so fat.

I see Roy Phillips, the man with the MYNAH licence plate, stride by, looking taller that I would have guessed from talking to him through the rolled-down window of his car. My sister and I stand at the field and watch the geese feed on the pitifully sparse green patches of winter rye, and I listen to people marvel at their numbers, count collars, juveniles.

Sean Boyd calls a greeting as he hurries by, finished with his stint at the scope, and ready to present another slide show and lecture in the meeting room across from the gift shop.

We continue on to the observation tower Henry Parker built, where we find five men in the fluorescent striped vests of volunteers standing on the platform at the top. Two scopes are set up there. One man, a volunteer originally from Montreal, whose toque is pulled low over his forehead and ears, to protect him from the wind that slices up from the marsh, says he came out here because he heard there was a bald eagle sitting on every snag on the foreshore, each with a snow goose for lunch. 'We get a lot of snow geese in Quebec but no eagles. I like to photograph events like that, but I don't have my camera with the telephoto lens.'

So has he seen what he came out to see, we ask? No, he admits, but there are two swans out there, and he angles the scope west, into the sun, to show us. Then we swing it back to look at a northern shrike on a tree-top; harriers, which used to be called marsh hawks, gliding over the land.

Back in the warming hut, five or six people are watching a video about snow geese, and at a table nearby, children – including my daughter Annie – are drawing and colouring pictures of snow geese for the art contest. Winners in each age group will receive a Snow Goose Festival poster, two bags of bird seed, and free admission to the sanctuary on their next visit. Volunteers hand out crayons and paper, and send each kid home with a colouring book called *Wildlife and Farms Together in the Fraser Delta.* It's a way of educating the public about the necessity for farms and wildlife to coexist, and features animal characters such as Henry the Heron and ducks who say things like, 'It's time for snackin' and quackin'.'

Outside, the cold air is infused with the smell of hot dogs. The food tent volunteers courteously inquire whether customers prefer their hot dogs with or without onions. There is also coffee and hot chocolate, and cookies, two for twenty-five cents. I wonder if snow goose jerky would be a mover here, or if people would find the very idea sacrilegious.

Across the parking lot, the Orphaned Wildlife Society has set up a tent in which a female red-tailed hawk rests on a perch and a winsome male barn owl perches on a volunteer's shoulder. The man beside the hawk explains that it was found with respiratory problems and can't survive on its own. Its beak is fearsome, but the gentle curiosity in its eyes charms me.

Back in town, the town of Ladner, next door to the Delta Museum and Archives, where antique decoys are on display, the owner of Uncle Herbert's Fish and Chips is swabbing out the ladies' room. When she calls, 'Just a minute, dear,' her accent so matches John Ireland's I have to ask if she too hails from Yorkshire and if she knows John. The answer is yes to both questions, yet she has never set foot in the Reifel Sanctuary, or watched the snow geese. 'But I hear it's somethin' else to see them all fly up, flappin' ...'

Geese in a Westham Island farm field

John Ireland has been resident manager of the sanctuary since April of 1984, and after more than ten years he impresses me as a steward in the sense that he seems to be responsible for the place not only because it is his job, but because he is, in the literal sense of the word, the care-taker. A former bricklayer, who did not study science in university because, he says, he didn't feel he had the brains for it, he's the sanctuary's all-around man. I've seen him hammering benches together, building observation blinds, scattering seed, scrubbing out washrooms, directing traffic, guiding groups of school kids – one of his favourite tasks – and leading the weekly ten o'clock Sunday morning tour around the sanctu-ary paths.

At forty-six, his hair and beard are grizzled with grey. Although he has been watching birds since he wandered into a museum in a park in his hometown of Huddersfield, Yorkshire, and became fascinated by the stuffed birds he saw there, and has since watched birds in most countries on the planet, it is still what he likes to do best. He often spends his day off down the road at Boundary Bay, observing birds that don't use the sanctuary.

I think of a passage I read in Konrad Lorenz's *King Solomon's Ring*: 'Every shepherd knows his sheep, and my daughter Agnes, at the age of five, knew each one of our many wild geese by their faces. Without hav-ing known all the jackdaws personally, it would have been impossible for me to learn the inner secrets of their social life. Have you, dear reader, the slightest idea how long one must watch a flock of thirty jackdaws and how much time one must spend in close contact with them in order to accomplish this end? It is only by living with animals that one can attain a real understanding of their ways.'

While John Ireland does not exactly live with snow geese, he prob-ably comes closer to living with them than most of the people I talked to. He and his wife Mary Taitt, a biologist who specializes in Townsend voles, their young son Timothy, a dog, and the two sandhill cranes that have taken up residence in their front yard, occupy a bungalow the Waterfowl Society built to house the sanctuary warden in 1967. From this choice spot John can hear the snow geese as they arrive each fall, the first group of one or two hundred, then a larger flock of, perhaps, a thousand,

and finally the whole throng, tens of thousands. This is the time of year most people visit the sanctuary, too: 2,000 to 3,000 a week from October to December.

Though the sanctuary is open year-round, and visitors, including Roger Tory Peterson and Prince Phillip, have dropped in at various times, there's more publicity for the sanctuary in the fall because of the Snow Goose Festival. Visitors from the city seem not to know that by the time the banners are strung and volunteers set up the various displays, the snow geese have been feeding on the marshes and fields for weeks, and will be here for weeks more: they seem to think the snow geese land here for just a few days or a week, as if they have come for the festival that honours them.

When I return to the sanctuary later in November, the geese are far out on the foreshore, almost out of sight. In closer range, a black-and-white blur of bufflehead ducks flit up from the slough. Mallards whir by. I count six or seven harriers gliding over the marsh. In the grey light, before the rain, the shore seems closer, the marsh beaten down, log strewn. The rain begins and continues on and on, it seems, for the rest of the month. To those of us who live in this climate the sound of rain can be soothing. When it stops the quiet has a waiting, unfinished quality.

By the 8th of December the Wrangel nesters that were only stopping here on their way to California have moved on; 20,000 have settled on the Skagit Delta, and 22,000 remain on the Fraser.

One day I find a field of geese behind the barn where Henry Parker kept his decoys. Some big adults definitely seem to be on patrol: white, healthy, their necks erect, their dark eyes bright. The backs of the young are mottled with the brownish-grey of their juvenile plumage and new clouds of white feathers that will gradually replace the grey feathers that mark them as juveniles. Rain needles down making glue of the fields. Mud clings to their webbed rose feet.

Squawking, honking, some quietly, as if they are engaged in mundane conversations, others louder, argumentative. One digs up something that looks like a carrot. A root?

A small hawk sitting in a treetop across the road, its wings slightly raised, is finally blown off the tree by the gusting winds. The needles of rain have thickened to lances. Out in the strait, the wind is so strong the evening news reports that the ferry from Vancouver Island was unable to dock.

ON THE FIRST DAY of the new year, after a week out of town, I dash out to Westham Island to see if the geese are still here, for they traditionally leave for the Skagit in early January. The parking lot is crammed today, every bit as crowded as the day of the Snow Goose Festival. And the geese are still here, at least 5,000, I find out later, from Sean Boyd. A large portion of them are feeding on a sandbar between Westham Island and Steveston, the historic fishing community on the north side of the river's mouth. Two thousand are feeding on the dykes around the Vancouver International Airport though, and this group will stay throughout the winter, the first year Sean Boyd has recorded such behaviour.

A couple of weeks later I make another trip out, to gather more notes from John Ireland. I notice wire down by the Westham Island bridge and find John's house and the sanctuary gift shop without power because another swan collided with the wires. This has been happening since electricity came to the island, according to Betty Husband. Blackouts are a way of life. On Christmas Day, says John, residents of the island didn't know if their turkeys were going to get roasted before the power went off again.

Bob Bradlee, BC Hydro's wildlife biologist, recorded six swans killed in December and January, and 7,000 customer-hours lost to power outages. Snow geese hit the wires less often, but one was reported killed in the '94-5 season. The trouble is that in certain light, sunrise and sunset, birds have difficulty seeing the wires. So later in the month, BC Hydro tried flagging two sections that have caused the most trouble. Blue, red, and yellow plastic fluttered in the wind over winter fallow fields, same as they do over used-car lots in the city. But this was just a temporary measure. Hydro's long-term plan was to install bird diverters in problem areas, fibreglass spirals around wires to make them more visible.

We can do without electrical power today, however, because there's enough sunlight coming through the window for me to be able to read the list of dates John has compiled, of snow goose arrivals and departures over more than ten seasons. He offers to show me the scrapbooks he and Varri Johnson, who runs the gift shop, have put together, and we talk for a while about the general impressions he has gathered of geese since he's been here. Although he is friends with several of the hunters I talked to, and they supply much of the food for the sanctuary's annual pig and corn roast, he doesn't like it that hunters are allowed to shoot right outside the sanctuary, and he thinks the season begins too soon.

'The thing that hits me is that hunting starts just as they're coming in. This beautiful flock of snow geese arrives and people start shootin' em. That isn't very nice, is it?'

He doesn't mind a good hunter who gets one or two because he recognizes that it's a tradition. 'It's the yahoo types that get to me. I've seen guys go into a field of 10,000 birds, start blasting. Kill ten, injure ten. It's sort of sickening.'

Though he is an amiable fellow, seemingly able to converse and joke with any of the wide variety of people he encounters on his job, his view of snow geese is that of a bird lover, and he does not gladly suffer those who complain about the presence of snow geese and swans in a place that is natural habitat for them. As for farmers shooting geese as a way of protecting crops, he says, 'Okay they're eating crops, but a dog would get rid of them. A lot of crops are better for being cropped.'

He has little more patience for the agencies that manage wildlife, complaining about jurisdictional misunderstandings, provincial officials who don't communicate with federal officials. Today he is particularly miffed about some harsh changes the CWS made to the farm at the sanctuary's entrance. Brambleberry thicket bulldozed off, the little pear tree that must have grown there for nearly a century destroyed. Though the Wildlife Service claimed it was cleaning up the area for safety reasons, John saw only the destruction of good songbird habitat.

'We pay them to protect these things, then we have to remind them that that's what they're supposed to be doing!'

Aldo Leopold wrote, 'Let no man jump to the conclusion that Babbitt must take his PhD in ecology before he can "see" his country.' John is a good example of someone who sees a lot because he is privileged to live among wild birds: his agenda is none other than that of caring for a place that is meant to be a refuge for them. And he claims he is just as happy to watch a robin or a flock of geese as one of the rarer species. He discovers much about their behaviour simply by being near them every day. At dusk, for example, he notices that the geese stretch their wings before they lift up and fly out to the foreshore where raccoons and coyotes cannot bother them. 'Who goes first?' he wonders. 'Is there a king goose, a queen goose? A couple?'

He has also noticed that the mates and young stick with an injured bird for a long time, a bird downed by a hunter's shot or injured after colliding with a wire.

Yet, with all that scientists have learned about snow geese and all the rest of us have observed, there is much about them that will remain mysterious, understanding denied us because we are a different species.

EARLY THE NEXT WEEK the rain is pouring down, as it is in most locations along the Pacific coast, as far as southern California, which has experienced flooding off and on for the better part of this new year. The ditches are 80 per cent full of water, the berry bush tangle above them still a dark Christmassy green, the branches glossy. Just over the Westham Island bridge, I stop on the road to look at the trumpeter swans mucking about in the potato field where I usually see them. Beyond the swans, almost out of range of my binoculars, an assembly of snows alternately root, rest, fly up. Directly in front of me, a juvenile bald eagle flanked by two mature bald eagles, perches in the centre of a semicircle made of four adult trumpeter swans. Three to six metres to their right is another odd mixture of the same species: two adult swans, a juvenile bald eagle, and two juvenile swans. The big, fluttery flock of snow geese seems to be moving southward, down the field behind the swans and eagles, towards a farmhouse that stands above Canoe Pass.

I hear a loud sound, someone beating on an oil drum perhaps, or firing off one of the exploders meant to chase birds out of fields. The geese swirl up, move farther away from the farmhouse. It may be the farmer trying to move the geese and swans off his fields, or an eagle that stirs them up. It is the most dramatic, most awe-inspiring sight of snow geese, several thousand springing into the air simultaneously, their

This foursome may be a family flying together.

distant pattern of moving white making a path across a sky that is, today, the colour of fish flesh. But the spectacle we so enjoy costs the birds a good deal of energy, for it is much harder to flap than to glide. In *How Birds Fly*, Anna Maria Caldara explains that when they flap, birds must produce thrust – the force that propels them forward – as well as lift. 'Lift and thrust are achieved by a flapping motion of the wings, which can be divided into an upstroke and a downstroke. When birds flap, their wings are engaged in a dual action, working not only as airfoils, but also as propellers.

'Two pairs of strong breast muscles create the up-down motion of the wings. The largest flight muscles are the pectoral muscles; when they are flexed, downstroke occurs. When the pectorals are relaxed, an upstroke is achieved by a pulled action of tendons over the joints of the shoulders.

'When taking off, a bird must be able to secure sufficient lift to get its weight off the ground in as little space as possible. A spring brings them high enough to begin a slow, but strenuous flight movement of their wings.'

Once it is airborne, a bird's wings need not work as hard, but as they prepare to land they must expend another big spurt of energy. Sometimes big birds make it easier on themselves by using the wind to slow themselves down, by landing into it.

'To slow down, birds lose speed by bringing their bodies into an almost vertical position and spreading their tails; the fanlike shape of the tail acts as an airbrake ... As it must do during normal flight, a bird that is slowing down must produce lift that is equal to the weight of its body. In order to accomplish this while travelling at slower and slower speeds, the bird must increase its angle of attack. At high angles of attack, however, the air does not flow smoothly and evenly over the wings. Rather, the air separates from the surface of the wings, creating turbulence over them. This uneven airflow severely reduces the amount of lift, and it will eventually cause the bird to stall.

'In order to prevent stalling during slow down, birds use a small group of feathers called the alula. Birds' wing tips, like human hands, have a row of finger bones. It is here that the flight feathers are situated.

The feathers of the alula are attached to the bone that corresponds to the thumb and are folded back out of the way during flight. But to prevent stalling, the alula is spread forward: air streams through the slots or spaces in the spread feathers, restoring an even, smooth air current over the wings. With the alula, the wing can provide lift even at very slow speeds.'

I think of kids playing airplane, arms extended straight, fingers angling up and down as they zoom over the countryside in their imaginations. Birds evolved from reptiles, about 150,000,000 years ago. And though we were never directly connected in the evolutionary tree, we share with them warm blood, a backbone, two pairs of limbs based on a pattern of five 'fingers,' and a love of flight, though in our case it is often the kind of flight that is possible only in dreams.

# *Pair Bonding on the Skagit*

IT TAKES AN HOUR AND A HALF for a snow goose to fly from the delta of the Fraser River to the delta of the Skagit River, and about the same amount of time to drive down, give or take a few minutes, depending on traffic at the border between Canada and the United States, something snow geese don't have to contend with.

Snow geese follow the coastline, but humans in cars must nose down the grey highway, Interstate 5, through Bellingham, past Samish Lake and up Bow Hill, then gradually descend into a classic scene: squares of winter brown soil, rectangles of new green, the river winding through, the town of Mount Vernon with signs here and there heralding the annual Skagit Valley Tulip Festival. For it is flowers that the local people publicly celebrate, and the 900 hectares of daffodils, tulips, and iris planted by bulb producers attract several hundred thousand visitors to their vivid fields each April. The tulip festival has, in fact, become such a lucrative tourist venture that someone suggested the valley also host a swan festival, in honour of the hundreds of trumpeter and tundra swans that winter around Skagit Bay.

Good idea? Maybe from the point of view of tourism promoters and birders, but Washington State's wildlife biologist for the area, Mike Davison, said that when farmers heard the suggestion, 'they went ballistic. They threatened to blockade the highway.' For odd though it may seem, tourists have become one of the more challenging wildlife management problems in the Skagit.

This has a lot to do with the geographical position of the valley, which lies almost exactly between the major metropolitan centres of Vancouver 121 kilometres to the north and Seattle, 97 kilometres south. The Skagit River itself actually rises in Canada, in the mountains of Manning Provincial Park, then flows through the Cascade Mountains, impeded for part of its length by the Ross and Diablo Dams, then for 125 kilometres through the county that took its name. The Skagit is the largest watershed in the Puget Sound basin, with 2,900 streams contributing over 20 per cent of the freshwater that flows into the sound. It is the second largest river in the western United States, the first being the Columbia, which flows beneath a big bridge on the interstate some 322 kilometres farther south. Just west of the town of Mount Vernon the river branches into the two distinct channels that cradle Fir Island, same as the branches of the Fraser hold Westham Island in British Columbia. Both rivers have, over time, deposited silt that has created productive farmland and watered the marshes at the rivers' mouths, encouraging the growth of the bulrush and sedges waterfowl like to eat.

In early October, when the big Wrangel flocks work their way down from the Arctic, a portion heads directly for this delta, where they will spend the whole of the winter. About half of the flock used to winter here, half on the Fraser, but in recent years the percentage has changed. Sean Boyd's population studies show that lately the mix has been about seventy/thirty, with the greater number staying on the Fraser until January, when the situation is reversed. From the point of view of birds that travel over 4,850 kilometres to reach their winter home, the Fraser and the Skagit may appear as a single extended pantry. But there are some important differences. For one thing, until 1995, only the Fraser provided an upland refuge the snow geese could use during hunting season. For another, there are fewer hunters in the Fraser, though, as I noted in Chapter 7, that does not mean fewer birds are taken every year. No one knows if hunting affects the choices the geese make, or if there is some other reason they have been distributing themselves slightly differently of late, but another change is coming that could increase this trend.

On the bulrush zones in the marshes around Skagit Bay, spartina, a saltwater-tolerant plant common in marshes along the eastern coast of North America, is spreading north and threatening to displace the native flora. The man who planted it, a farmer and an economist who owns three-quarters of the estuary of the Stillaguamish River at Port Susan Bay, the next bay south of the Skagit, wanted to try to graze cattle on the marsh, on this spartina. What's happened as a result is typical of the ecological changes that take place when a non-native plant is introduced. Spartina appears to be out-competing the nutrient-rich three-square bulrush snow geese prefer. No one knows if the geese will eat spartina, or if they do, how useful it will be for them. But if it continues to spread, which seems inevitable, it could have a serious effect on snow goose habitat in the Skagit, and the redistribution of the Wrangel flocks could become even more dramatic.

If so, it would not be the first time snow geese changed their use of marsh habitats along the coastline of British Columbia and Washington. A Fir Island dairy farmer, Maynard Axelson, recalls his father and some of the Fir Island old-timers talking about the season they saw snow geese for the first time. 'All of a sudden there were all these white birds and we shot a bunch of them and we weren't even sure what they were,' his dad told him. Maynard and others suspect that the geese began using the Skagit flats and the agricultural fields adjacent to the Bay only in about the 1920s or '30s, perhaps because more and more people were beginning to move onto the Fraser Delta. There certainly is proof that snow geese have been landing on the Fraser for hundreds and thousands of years: snow goose bones were found in a burial pit in Boundary Bay that archaeologists dated back 3,500 years. And sailors who explored the Fraser in the early 1800s reported seeing numerous white geese at the river's mouth.

On the other hand, the flocks that prefer the Skagit may have moved there from habitat farther south. Suggestions of that possibility come from stories of Coast Salish people recorded by the anthropologist

William Elmendorf. In *Twana Narratives*, Henry Allen tells of the white geese that came in big flocks to the Skokomish Flats, the estuary of the Skokomish River that drains into Hood Canal at the southern end of the Olympic Peninsula, about 129 kilometres southwest of Skagit Bay.

Native people traditionally hunted birds with small bows and arrows fitted with prongs sometimes made of swan bone. They also caught birds at dusk in nets stretched between two vertical poles. And they sometimes shouted loudly, trying to alarm low-flying flocks so they would fly into the nets. The artist Paul Kane saw such nets in use as late as 1859 on the shores of the Strait of Juan de Fuca.

The small canoe they used for waterfowl hunting was sharply cut away at both prow and stern and was large enough for only one person, like duck punts used by non-Native hunters today. Native hunters would canoe out into the bay at night, with small fires burning in a hearth in the bow of the canoe. This was called jacklighting.

Jacklighting was a hunting practice employed by many bands along the northwest coast. In his *Waterfowl on a Pacific Estuary* Barry Leach included an eyewitness account from an earlier book of a brant hunt off Vancouver Island: '"A dark wet still night is chosen in the winter, when the Geese are feeding on the beds of Zostera (eel grass) in shallow water. Two Indians go out in a canoe, one in the bow armed with a torch of resinous pine splinters known as a gum stick, and a large net like a landing net on a pole; the other sits in the stern and paddles the canoe in the direction of a flock of Brant. As soon as the canoe has got in amongst a flock the torch is suddenly lighted and as suddenly extinguished. The birds at once get up and fly about a short distance but settle again as soon as the light disappears. The Indians mark the direction taken by the birds and follow them, again paddling noiselessly into the flock. The torch is again lighted and extinguished with the same result. After this manoeuvre has been repeated some three times the geese become bewildered. When the torch is lighted they do not attempt to fly but stay and gaze at it. They are then quickly scooped out of the water by the Indian with the net." (Sprot, 1928).'

Although waterfowl did not compose a large percentage of their diet, the seasonal hunting of migrating birds was part of Coast Salish life.

In addition to eating the meat, Salish people wove the down of geese into blankets and used it for ceremonial purposes. One account of a secret society initiation describes people dressed in coats of white feathers, with their heads also covered in white feathers, so that just their faces showed. In another story, white goose down was sprinkled over a Neah Bay woman to make her invisible.

In this culture spirits exercised powerful effects on people's lives. In *Twana Narratives*, Frank Allen told of how one time in the mid-1800s, when the geese were going north, the Skokomish people caught lots of them, and gave a feast and invited the Duhlelap headman and his large family to come. 'They fed them lots of geese, but because they didn't give them any to take home, [the headman] got mad. And when he got home he got to feeling more and more hurt. So he told his folks, "I'm mad about that trip! I'm going to fix those [people]. I'm going to send my sad'a'da" – his spirit power – "up there to them." So they had a ceremony and he sent his sad'a'da to where the geese land. Lots of them. Where everyone comes to get geese. That's near the mouth of the Skokomish. There was good hunting for geese at night, with jacklights. Well he made his sad'a'da stand ... with its arms held out wide. That power there acted like a scarecrow. The geese got scared and wouldn't land there and flew on.

'In my time I have heard people say, "Oh we used to have lots of geese, but for a long time they've been getting less every year." That was a dirty trick of [the headman]. His sad'a'da is still there driving them away.'

It's interesting to speculate whether or not the reported disappearance of the geese from the Skokomish Flats was a reaction to the inhospitable behaviour of some Skokomish people, or if this was just their way of explaining why geese no longer showed up because, perhaps, they had exhausted their food supply and moved north to the Skagit or the Fraser. But another story appears to reflect the seasonal movement of the geese from the Fraser southward. 'It was winter at this time, but that Klallam man said, "Soon it will be here, food will be here on the Skokomish Flats. White geese [qa'wqaw'] will be here." And in a day or two all kinds of those geese showed up on the flats. And people killed lots of them from canoes at night, jacklighting. And that man had them bring him those geese, and everybody there ate them with him.'

THE WRANGEL SNOW GEESE still stop in various fields and marshes on their way to California, and a flock of about 1,000 winters on Sauvie Island in the Columbia River near Portland, Oregon. Then just after Christmas, the section of the flock that spent the fall months on the Fraser begins drifting down, until, by mid-January, almost without fail, the entire northern portion of the Wrangel nesters have settled around Skagit Bay and Port Susan Bay, at Stanwood. Although some people believe that individual geese and family groups move back and forth from the Fraser to the Skagit on a more or less continuous basis, Sean Boyd and Barbara Pohl have been studying the inter-delta exchange and feel certain that there is a group that prefers the Skagit in the fall, and a group that prefers the Fraser, and that, until January, they remain fairly discrete. The southward movement of the Fraser flock coincides with the end of hunting season on the Skagit, and this may have something to do with the timing, but temperature also plays a part. Long-time residents say that weather whistles right down the Fraser and can keep the northern delta two to three degrees colder on average in January. Whether it is because the snow geese feel safer or like the slightly milder weather, they ensconce themselves here firmly every year, even the year the dykes around Fir Island broke and all the farmland flooded. There were two big floods in the last decade, but the worst was in 1990. Madge Axelson, Maynard's mother, showed me the high-water marks on the grandfather clock in her living room, and pictures of Maynard carrying her on his back out the front door to a boat.

Maynard, who has had what he describes as 'a real intense interest' in birds since he was a kid, used to keep various species of ducks and geese, including snow geese, pinioned and penned in an enclosure behind the farmhouse. Groups of school kids would bus out to the farm to visit his aviary. But he lost all his captive geese in the flood: they floated away or were taken by coyotes or some other predator. Now he has only a few grey brant in a pen behind the farmhouse, and his dairy farm keeps him so busy he doubts he will start another collection of captive birds. His plans are to open his barns for tours of the dairy farms instead.

This will supplement farm income, he hopes, for while he milks seventy cows and also sells bulls, heifers, and cows, the price of milk is about the same as it was when he took over management of the farm nearly twenty-five years ago. Maynard's grandfather established the Axelson's Fir Island farm in 1903, and though, because of his interest in wild birds, Maynard considered wildlife biology as a career, his father's heart condition forced him to make a choice and he opted to continue the family farm. In his mid-forties now, he studies bird behaviour as an amateur observer and photographs birds, and he has travelled to Alaska five times as a volunteer on banding expeditions. We sit in the modernized kitchen of the Axelson farmhouse, where the cow motif marks tea towels, fridge magnets, photographs, china knick-knacks, and he sets up a light table alongside stacks of photograph albums and boxes of slides, scores of them of snow geese. 'You asked for it,' he jokes.

I had read that snow geese choose their mates in the winter and spring, long before they actually mate. In his *Handbook of Waterfowl Behavior*, Paul Johnsgard describes various rituals: 'The male courts the female by swimming ahead of her, head held high, tail somewhat cocked. The male attacks any intruder, first stretching its neck up and spreading his wings, then flicking his wings at intruding birds. He then returns to the female, head held high, and they stand next to each other clamoring.' While Johnsgard acknowledges that this would seem a strange way for humans to 'marry,' it works to establish a bond for life between snow geese. But he also describes a chase, where the female takes off pursued by more than one male. The male sometimes pulls at the female's tail.

The scientists I talked to had not gathered much specific information about the pair bonding of Wrangel snow geese, but as it was supposed to take place in late winter, just before they began the move north, it seemed the geese had to be paring up on the Skagit. Maynard confirmed it.

'It goes on all the time. You can see it here any day of the week. I just saw it here a couple of days ago. A bird will be sitting in the field and you'll see these small groups of birds, like three, and they're doing these wild figure eights and circles in the air. What it is is two males chasing a female, trying to pair up with her. Oh yeah you see that all the time. And if you watch them close on the ground, you'll see a pair and the male will herd the female, chase other males away.

'Of course I don't have the scientific evidence to back this up. I mean, it's just visual observation. It's just like the ducks this time of year, you know. Everywhere you look they're in pairs now. But the geese especially will do this aerial thing I'm talkin' about, you'll see that a lot. I just saw it a few days ago.'

Whether or not these chases are birds actually pairing, or a male defending a female he has already paired with, as some scientists now believe, the geese involved are likely to be a mixture of many age groups. This is because Wrangel Island does not produce geese on an even yearly basis. Because some years – as in 1994 – the young made up only a small percentage of the flock, whereas in '93 they made up 40 per cent, there may not be enough birds in the same age group for all to find mates. Another factor that complicates the mix is the fact that pairs are broken up by hunters, and some too by flying into power lines or falling prey to coyotes or raccoons.

In a paper comparing the Wrangel Island nesters with the snow geese that nest at La Pérouse Bay, the Russian scientist Syroechkovsky wrote: 'As is well known, the geese form pair-bonds on the wintering grounds. And apparently young geese extremely often form pair-bonds with older individuals. This may be caused on the one hand by the fact that usually the number of young birds on the wintering grounds is comparatively low. Geese are long-lived birds and the proportion of each cohort is small. On the other hand, rather many mature birds on the wintering grounds have lost their partners for various reasons and more actively search for a new one than do young geese. Thus, apparently in the formation of pair-bonds, a significant proportion of young females form bonds with old males, and young males often unite with old females.' On Wrangel scientists have seen two-year-old males accompanied by females with last year's offspring.

Generally speaking, young birds have a long betrothal period. Barbara Ganter described it this way: 'They pair bond in the winter or spring. The birds somehow get together and form a pair and run around together and behave together and not necessarily copulate but they really form a pair. It's just like people.'

Konrad Lorenz theorized in *King Solomon's Ring* that wild geese had a long 'engagement' period prior to actually mating as a way of strengthening the bond that was necessary to keep them paired throughout their natural lives. 'Amongst those few birds which maintain a lasting conjugal state, and whose behaviour in this respect has been explored to the very last detail, the betrothal nearly always precedes the physical union by quite a long period of time ... Jackdaws, like wild geese, become betrothed in the spring following their birth, but neither species become sexually mature till twelve months later, thus the normal period of betrothal is exactly a year.'

Scientists have since discovered that birds pair early because of competition among males for the best quality females, which leads to pair formation before it is biologically necessary.

Maynard has no photographs of snow geese engaged in a pair-bonding ritual, but he has pictures of them doing just about everything else. One dramatic shot shows a flock of tens of thousands silhouetted against an orange-black sunset sky. Another features the rare, to this part of the world, blue phase of the snow goose, a single 'blue' snow goose that showed up on Fir Island three years running. It might have been a captive bird that got loose, or one that drifted over to the Pacific flyway from the central or eastern flyway, where blues are more common. Other pictures illustrate the way snow geese cluster in family groups through the fall and winter. Maynard used these to show how snow geese depend on farm fields in the Skagit Valley. The acreage devoted to pasture has dwindled as dairy farmers lose heart at low prices and family farms disappear. So Maynard, a committed dairyman, and a lifelong wild bird enthusiast, helped promote the 'Barley for Birds' program developed by the State Fish and Wildlife Department, Ducks Unlimited, and the US Fish and Wildlife Service.

The day I spent with him, Mike Davison showed me a plot of barley where he saw snow geese and swans feeding on grain for the first time in 1993. 'Barley normally takes 120 to 180 days to grow, but this strain grows in sixty days, so farmers who harvest an early crop could put this type of barley in for a second crop. Because the barley kept the weeds down, it meant fewer herbicides had to be used, and this particular type of barley also picked up excess nitrate in the soil deposited by commercial

Barley for Birds program on Skagit Delta

Snow geese and cover crop on Westham Island

fertilizers. But the best part of it is that the snow geese ate it.' Before then they foraged on grasses and waste potatoes and sometimes cucumbers left to rot in the fields around Skagit Bay.

Maynard Axelson's role in the program was to persuade local farmers to offer plots to experiment with, and their cooperation demonstrates the difference between the way snow geese are looked on by farmers in the Skagit as opposed to farmers in the Fraser Delta. I recall the December day I stood on the side of a road on Westham Island watching geese with Barbara Ganter and Barbara Pohl. A woman drove up to us slowly, rolled down her window and called: 'Remember who feeds them!'

She was the wife of a farmer, I'm sure, expressing the difficulty that sometime arises from the agricultural/wildlife land-use conflict. Henry Parker and Robert Husband had complained about the damage swans do to a field, their big feet packing the soil so densely that, Henry claimed, one farmer's tractor had stalled trying to plough up the field for spring planting. Sean Boyd told a story about a farmer out on Brunswick Point who, frustrated at the presence of thousands of geese in his field, had gone out with his shotgun and blasted ten birds in a matter of minutes. The Canadian Wildlife Service and Ducks Unlimited Canada had devoted a whole colouring book and many information campaigns to the message that wildlife and farmers can get along.

But in the Skagit, apparently, there is no conflict between agricultural and wildlife use of delta lands. Mike Davison thinks that may be because so much more land is devoted to agriculture in the Skagit. In the Fraser Delta, bedroom communities are spreading farther and farther into traditional farmland, putting a strain on both farmland and wildlife habitat. In the Skagit, if one farmer chases geese out of his fields, the flock has hundreds of other fields to settle in. There's a lot more land, and a lot more waste crops for both snow geese and swans, says Maynard Axelson. 'They can go through and root for potatoes and carrots, there's a lot more corn in the fields. And there are fewer winter crops raised here.' So Skagit Valley farmers don't consider waterfowl pests.

The pests are the people who swarm here to watch snow geese and swans, who pay little attention to the fact that farms are private property and that snow geese waste hard-won energy when they fly up to escape the approach of a human intruder.

'We're in the backyard of King and Snohomish counties, Seattle, Tacoma. And this valley is so intensively advertised by the media that we're just inundated by people coming here to see the swans and the snow geese,' Mike Davison explains. When snow geese have been feeding in the same field for more than a couple of days, 'it is nothing to see 700-800 vehicles parked alongside the field on the weekend. If a flock is being flushed thirty to forty times a day it actually does biological damage to the species.'

The behaviour of some of the visitors bothers Maynard Axelson, too. 'What really irritates me are the drivers of big trucks who lay on their horns trying to get the birds to fly. Just time after time, it's so stupid! I guess the idea is to get them to fly so you can enjoy watching them. But if you enjoy watching them, why scare them out of the field? It's self-defeating.'

Some photographers, more concerned about the perfect shot than the welfare of the geese, actually walk out into a field for a closer view, or to stir the flock up into the spectacle that makes such a thrilling sight, such a dramatic photograph. This not only affects the birds, it annoys farmers who take umbrage at the gall of people who do not seem to care that they're trespassing on private property.

I'm learning all this as I ride with Mike Davison around Skagit Bay on an unseasonably warm and radiant day in early February. His mission is to monitor those radio transmitters that were meant for California birds but ended up instead on some of the Skagit-Fraser portion of the Wrangel snow goose population. There are 40,000 geese around somewhere, but we have no luck at the first place we stop, a gorgeous and rare piece of intact woodland leading to a bluff that overlooks the north end of Skagit Bay. A small group of rusty-hided cattle saunter up, curious to know what we're doing here, it seems, but we see no snow geese, and when he scans through the various bands on his radio, Mike hears no answering chirps. And so we drive on, back out the mud and gravel track that connects this stretch of private property to the paved public road. As the area wildlife biologist, Mike has access to private lands in most of four counties, he says, his voice betraying both the pride he takes in this privilege and the responsibility he feels.

He describes his view of snow geese as that of a manager's though he trained as a biologist and oversees numerous scientific research projects as part of his job. Whether he was born to manage or simply grew into the job, there is as much natural husbandman as bureaucrat to Mike Davison, or so it seems to me. A few years older than Maynard Axelson, a man of medium height whose short brown hair is covered by a dark green cap emblazoned with the seal of the Washington State Fish and Wildlife Department, he was born in Mount Vernon and grew up playing with farm kids, hunting and fishing. 'When we played hooky, it wasn't on a nice day like this, but when it was rainy and windy and we could go hunting.' His father actually worked as a game warden in Germany during the occupation after the war, and the stories he told of his experiences there, along with the fishing trips he took his young son on so appealed to Mike that he decided at the age of five what he wanted to be when he grew up. He studied at Eastern Washington State University in Pullman, because that was one of the only places in the state that had a wildlife biology program at the time, and he has worked for the Washington State Fish and Wildlife Department in various areas around the state since the early '70s, but in his home area of Mount Vernon since the early '80s.

We drive onto another bayfront property, owned by a man who is a commercial fisherman, and his ease with the people in the district is apparent as the landowner comes up to the truck and he and Mike jaw a bit about fishing and hunting. It's spring and the person in question is going up to Canada to hunt, because spring hunting is not allowed in Washington, while it is still legal in British Columbia until 10 March, something that publicly incenses Mike. 'We are outraged,' he says, as we drive back out to the road. 'It's absolutely biologically, scientifically, and management-wise unacceptable. You ethically don't hunt birds in the springtime because that's the breeding season. You're not just shooting birds, you're shooting bonded pairs. You shoot one, you're shooting two. It's not accepted on any level anywhere in the world, and yet we see brant being shot in March, snow geese being shot in March.'

Mike says his opinions are well known to the management agencies in Canada, which justify the spring hunt on the Fraser because it's a tradition, and a way of controlling crop losses to farmers. Because the Fraser is out of his jurisdiction, there's nothing more he can do about it, but his view remains uncompromising. 'You don't hunt birds during the breeding season, you just don't.'

It's the view of a steward as opposed to a farmer concerned about growing plants, or a Native person for whom geese in the spring have been a traditional source of food at a difficult time of year, but Mike Davison makes no apologies: it is his job to look after wildlife, the 'resource.' He refers to the birds and mammals that populate his jurisdiction impersonally as 'the resource,' and he is quick to draw a line between his responses as a person and as an official wildlife manager. He even responds formally when I ask him about his favourite species. Favourite is an emotional description, he reminds me, but he does finally admit, with a smile, that 'As a person there's nothing like seeing a seven-point elk surrounded by a herd of thirty other animals in an alpine meadow, with calves. And I have a fondness for river otter, because they're so playful.'

His protectiveness on behalf of wildlife applies also to visitors, birders, a few of which we encounter as he drives up to one of nine public-access areas around the Skagit and moves aside a wire fence so that he can drive onto the dyke itself. Before proceeding in the truck, he walks up to tell them what he is about to do, so they don't become alarmed at his approach, think he is disturbing the birds.

I've been to this spot before, a small curving section of marsh where driftwood is pushed up against the dyke and bulrush and lighter-coloured clumps of invading spartina merge with the muddy shore. Here we see the first snow geese of the day, a group of about 4,000, a half a dozen wearing the distinctive red collars with their alphanumeric code. As Mike turns his radio receiver on and I open the door to step out, he warns me not to slam the door behind me or go too close to the geese.

By this time in my research I have spent scores of hours prowling the fields of Westham Island. I have sat on rainy days inside my car, pulled over to the side of the road, where the geese were feeding a few

Snow goose with band and radio transmitter

feet away in a sodden field. I've seen people walking, talking, exclaiming at about the same distance and never have I noticed the geese respond to my presence or the presence of other people. Not on the Fraser at any rate. Mike's comment reminds me of the careful approach I made back when I was first getting to know the geese. On the one hand I find it surprising; on the other it reminds me of how views are changed by place and time and personal experience. On the Skagit, geese may be subject to human disturbance more often than they are on the Fraser.

The day we talked, the state of Washington was about to close a deal on what Mike described as a large piece of real estate that has historically received a lot of use by snow geese, a 243-hectare (600-acre) piece of pastureland on Fir Island, just down the road from the Axelson farm. The plan is to establish a refuge and build a viewing tower, so that people can stand and watch snow geese without disturbing them.

Until then, it is possible to see them from the curving marsh that borders Fir Island, such as the birders on the dyke are doing while Mike listens for the chirp of a radio collar. The small flock is actively feeding, the heads of many geese totally immersed in mud. The young are whiter, having shed most of the grey feathers that mark them as juveniles, and they are relatively quiet for snow geese, seemingly content to feed. The Olympic Mountains in the background make a jagged plum line against a sky thick as heavy blue cream.

Mike has not picked up any signals from this group, but there should be tens of thousands more geese to check somewhere around here. We drive out the access road and head south, and as we travel along the nearly trafficless rural roads bordered by the simple geometry of farmland, the subject of hunting returns. Mike's a hunter himself, in fact as we prepared to set out on this drive, he joked that the snow goose decoys piled with nets at the entrance to his garage were the survivors of a fire in an outbuilding on the property where he lives in the heart of tulip land. He went out once this past season, and shot two geese. The daily limit in the Skagit is three, but he shot two because he could only eat two, he said. I wonder aloud how someone who hunts and deals with hunters can be concerned about birds being disturbed by human bird watchers. Where I come from, the public perception is exactly the opposite, that it's the hunters who are doing the damage, the birders who are making nonconsumptive use of the resource, as the bureaucratic jargon goes. He's an experienced speaker, someone who talks to groups as often as a hundred times in a year, and this observation of mine is all that's necessary to nudge him into a commentary on the big picture. He shifts his eyes back and forth from the road as he speaks.

In Washington State, dollars from hunting and fishing licences go directly into what he calls the management base, the fund that is dipped into to purchase areas such as the one that will become an upland refuge for snow geese and swans. With the pressure on waterfowl habitat, and particularly estuaries, being of such great concern, habitat protection is crucial. 'People who complain about hunters don't understand the role they play. There are some realities a great many nonhunting individuals aren't comfortable with, and one of those is that things die, that mother nature is probably the harshest and cruellest of all managers, if you will. And I don't think nonhunting types want to acknowledge that. That's a part of mother nature that they just don't want to deal with. And unfortunately it's because of that that the nonhunting people do so much damage to the management of wildlife. They don't understand the role they play. In managing wildlife, whether it be waterfowl or anything else, some of our biggest problems have to do with the political naïveté of nonhunting individuals. Management of wildlife species is just that. There is a normal mortality and you tap into that; there's a portion of the population that's going to die anyway, and there's a balance between habitat and population. These are normal concepts. Nonhunting individuals don't want to acknowledge that those animals die and that if properly managed hunting doesn't exceed that level that normally would not be there. So it's a matter of personal preference and tolerance, but by not being realistic and being rigid in their thought process regarding hunting, a great many nonhunting types damage wildlife, because they get involved in political issues that affect funding for management. Audubon quite frankly is one of the more tuned-in environmental groups. They don't come out here and attack hunting and hunters because they realize that historically hunters have bought and paid for, and paid for the management of, all our best wildlife areas. Until recently, bird watchers haven't contributed anything at all. And yet we've seen nonhunters and anti-hunters assault the people who are paying the bills.

'It's one thing not to hunt yourself and one thing not to approve of hunting, but if you are not politically astute enough to understand what role hunting and hunters play in the welfare of birds and wildlife, you can just as effectively kill populations and generations of birds by destroying our habitat by interfering politically. It takes more than just a philosophy to be a positive contributor to wildlife management. And a lot of the nonhunting people are very naïve about that. I have no problem with hunting and understanding the population dynamics of wildlife species and the role hunting plays in that, and understanding that I as a hunter have shared in taking the burden of management where nonhunters haven't for generations and generations. I'm quite proud of that. When nonhunting populations can step over the dyke and say they have contributed as much to the welfare of today's wildlife as those hunters they criticize, God bless 'em. They've got a long way to go. My message to the nonhunters, who I'm accountable for – you know I have to work with those folks as well – is, become educated, understand the issues, know what role you and hunters play. And understand also when you and hunters are fighting amongst one another, the individuals who would exploit the habitat and our natural resources by development and pollution and other ways, they win because we need your help in fighting those issues. Wildlife loses again.

'We have a very highly aware group of environmental organizations here. They understand all that. There are people who go around and naïvely criticize hunting but as a whole these organizations are very tuned in and understand that if they tip the balance in any way, wildlife will pay the price. They also understand that in this valley, because we have so many hunters and it is tradition – a great many of these landowners you see around here are hunters – if you come out and criticize the hunters too aggressively you are in effect criticizing the people who are supporting the wildlife.

'People who come from Seattle, who didn't grow up in that time as part of their lifestyle, they're the ones who go into deep shock when they come up and see hunting activity.'

I tell him about the meeting I attended, where hunters seem to have been under the gun, so to speak, of animal rights activists. 'I've attended some of those meetings up there and I believe there's more of an undertone of naïveté in some of the environmental groups up there as to the role hunters play and the political realities of wildlife management than in this area. Now I wouldn't say the same thing about Seattle. Seattle and Vancouver probably have more in common than the land in between them.'

We have been driving south, along the road that skirts Puget Sound, from Fir Island towards the town of Stanwood. We climb a hill above the bay, and Mike pulls in at a farmhouse that overlooks the Big Ditch access area. Below us, on the edge of the bay, I see broad streaks of white. 'There's the bulk of the birds right there. Ninety-five per cent of them.' But they are too far away for us to get a good look, or for Mike to pick up any radio signals successfully and so we drive back down the hill, onto the Big Ditch access road to the south fork of the Skagit River.

I know this place from a visit a couple of years ago, the first time I actively traced the movements of the geese from north to south. It was later in February that year, but colder. On my way to Camano Island, from the freeway I spotted a group of birders investigating ducks and gulls on an overflow pond south of Stanwood and stopped to ask where I might look for snow geese in the area. The group had observed a huge flock earlier that morning and the leader gave me directions to the Big Ditch. On my way there, over a rise above the white-fenced farm where I sat with Mike today, I saw a huge puddle of them in a field, drove down the hill and backtracked on a road to get a better look, and the flock rose, spooked by a small plane, I think. It rose and rose and rose, filled the horizon, making a fluttery white line between the dark blue of the hills of Whidbey and Camano Islands and the lighter blue of the sky. I drove between posted fields, to the end of the road, where I parked and walked an informal trail through bare brush to the mudflats. And there they were. The birders at Stanwood said the flock consisted of 7,000 or 8,000. There were certainly that many, probably more. I thought of the Coliseum in Vancouver, where the Vancouver Canucks used to play hockey. The Coliseum filled with people. The image of 8,000 geese sitting on bleachers made me smile.

The flock was gregarious that day, konking musically, busily. When an eagle cruised a section of geese, their voices reached a more anxious pitch and they skittered up, in groups of ten to twenty – sevens, eights, teens.

I sat on a relatively dry hummock, sun warming my shoulders. The geese fed casually, dipping under the water. But there was not much feeding going on. Most of them were resting, loafing, many with their heads tucked into the base of their wings. Others remained alert, neck up: the sentries.

Eagles and hawks toured the flock periodically, then returned to stumps to watch. The bald eagle was the most visually impressive – a white-headed black-caped flyer pursuing black-winged white geese.

The scene was groaningly beautiful. In the foreground, the tawny ragged remnants of last year's weeds, scattered driftwood limbs, brown and bleached-white logs. The brown-blue gleam of the Big Ditch under a veil of ice. On its bank, tissue squares of ice whispered and cracked as they loosened and thawed. The ice and watery patches on the mudflats, the ochre hummocks of weeds reminded me of the tundra where the geese would moult and raise their young in a few months' time. Then the chattery ruffle of white geese folded into the steely blue bay, calm as slate until, out from shore, it became progressively more corrugated. In the mid-distance the blue-green mounds of Camano and Whidbey islands. Behind and above them the white-blue snowy ridges and peaks of the Olympic range.

When the flock sprang up and moved across the sky, individual geese looked like filings controlled by an invisible magnet. I sat watching them for a couple of hours, until the wind picked up, reminding me that it was still winter. I drove back out to the highway and onto Fir Island, where I stopped at the Snow Goose Produce stand to ask where else I might find the geese. A clerk pointed me in the direction of one of the dykes, where I found a sign that showed the difference between the protected swan, and the snow goose. That sign, and the comment of a local birder, who said birders in the Skagit pay little attention to snow geese, because they are so common, led me to think that snow geese were not valued here.

But that was before I met people like Maynard Axelson who has lived with both the wild snow geese that feed on his pastures after season, meaning hunting season, and the captive birds he used to raise as a hobby. His father taught him how to hunt and he shot his first snow goose when he was in the eighth grade, and when he learned taxidermy, as a teenager, he practised it on birds he shot himself. Snow geese in various positions are mounted on stands he shows me in the rambling upstairs of the original farmhouse. He takes one down from the wall to show how he inserted wire into the wings to make them stand out. Close up I see how saw-like the bill is, its potential to inflict damage on plants and human skin.

Keeping snow geese in captivity showed Maynard a side of the species that did not particularly endear them to him. They were so aggressive at mating time he almost had to separate one nesting pair from another. They dug down into the earth to rip up grass by the roots, to chew through fence posts. He shakes his curly head. Though he enjoys watching them, snow geese are not his favourite species: he didn't like their aggressive behaviour. He prefers the smaller, shyer brant, who have, in his opinion, the most luxurious of any goose down. But he doesn't end the discussion there. Recalling the lengths the snow geese must go to survive the rigours of a life cycle that starts in extremely inhospitable conditions, he theorizes that the birds may have to be aggressive to survive. Aggressive nesting, determined feeding may have evolved as a way of ensuring that they can cope with a nesting habitat virtually non-existent some years because ice and snow refuse to give way; wintering grounds where the most nutritious food lies buried between an inch or more of thick delta muck.

'In the literature you read how the high arctic nesters sometimes lay three to four days after they get there. Well they have to develop the eggs, mate with their mate on the way up there. They can't wait until they get there to do all that. It's a tough job. They have to eat and avoid predators and travel 2,000 miles while they're doin' this. It's a hell of a job, you know.'

He acknowledges, too, that the destruction his captive snow geese caused in the enclosure he built for them, tearing up vegetation, destroying fence posts, chasing other birds, was partly the result of them being enclosed.

Because his captive birds became sexually mature, ready to nest, in a climate far warmer than where the wild birds nest, they made nests and laid their eggs a good two months earlier than their kind in the wild, on Wrangel Island. Maynard observed them 'bagging down' as he described it, a good three weeks before they were ready to lay. He shows me a picture of a goose he collared just to see how the goose would react to the collar. 'See, this is springtime, about February, March. See the females where they start to swell right here? That's called baggin' down. You can tell the males from the females just by looking right here because they swell up in that area.

'They get real heavy in the rear end. You can tell when a goose is full of eggs. She gets to the point where it looks like she's goin' to explode, and she's layin' within a day or two. And then sometimes they'll bag down like that and get real heavy and ready to go and then if the weather changes or they get disturbed or a raccoon gets into the pen and freaks them out, then they'll just suck them back up, they won't lay them. They must just dissolve or something. Once in a while a goose will get so far and she'll get so heavy like that and they get something they call egg bound and they usually die from it. And I don't know if it's that the eggs develop too far and they get bottled up in there or what. Like I say, this is farmer terminology so you're going to have to bear with me. I don't have a scientific study to back this up, I'm just tellin' you what I think, or what I observe. I have a photo somewhere of a goose in captivity who's bagged down so far her butt is draggin' on the ground.'

Maynard noticed behaviour in his captive geese that was similar to that Fred Cooke's team observed at La Pérouse Bay. The mothering behaviour of older and younger geese, for example. 'Older mothers are definitely more protective, tighter sitters. At first you can't see the nest, then on the fifth or sixth day, when it gets to where there's five eggs in the nest, she covers it with a tremendous wad of down. Covers it just beautifully. Just like an artist.'

The movements of the wild geese have changed in Maynard's lifetime as agricultural land use has changed. There's also a difference he has noticed in the behaviour of the flock depending on how many young there are. When there are lots of young, the geese come into the fields even during hunting season. In years when there's a poor hatch, such as 1994, they stay out in the bay longer, perhaps because the feed lasts longer, he theorizes.

In February, the birds that are not involved in pairing are lazy, resting a lot, feeding. But as spring approaches 'They act like a football team in its hurry-up offence. They'll sweep across the field, and do this pigeon hopping thing, where birds in the back will come out to the front, and they'll just keep moving. It's just like a flock of sheep moving across field after field. And they'll be moving all day. And they'll be so much more flighty. The whole flock will pick up, and maybe go out to the bay for a drink of water, and then an hour or so later, they'll all be back in the field. I mean it's just completely different. You can tell they're getting geared up for migration. When they're getting ready to go, they'll all stand up and start walking, their voices at an elevated level. When it comes just about time for them to leave, they never sit still. How many fields can they cover in a day?'

OF THE 48,158 BIRDS Sean Boyd counted in November, 41,131 – including the 1,000 that spent most of the winter on Sauvie Island in the Columbia River – remained in the Skagit on 9 February. For the first time since Sean Boyd has been keeping records, a small flock of 2,000 never left the Richmond dykes to join their compatriots in the Skagit, but stayed in the Fraser Delta.

# Spring, the Flight North

By MARCH, the northward migration has begun. The sign on the window of the Reifel Sanctuary notes the return of snow geese to Westham Island the first week of March, which is generally consistent with the date of return John Ireland has recorded since he began keeping notes on bird movements at the sanctuary in the mid-1980s. It starts with a few, sub-flocks on the move from the Skagit. Not the whole flock, for in the first week of March thousands are still grazing their way across the farmland that borders Skagit Bay. But a few thousand have made their way up the coast to the Fraser Delta where farmers are readying their fields for spring planting. It's the yellow time: forsythia explosions along narrow farm lanes, dandelions springing out of ditches, clumps of daffodils everywhere. Big bunches, planted by Varri Johnson's father, butter the slopes above the sloughs at the Reifel Sanctuary.

Winter brown is merging with spring green, in the moss that furs sodden fields, in the sedge shoots poking up from mud bars in the river where, in the middle of the month, I see snow geese feeding. This is the habitat they prefer in spring, the mud and sandbars that rise like the knuckles of fingers submerged in the estuary, and slightly upriver on the Ladner marshes.

But they use farm fields around the sanctuary occasionally too, and John Ireland has seen them engaged in the pair-bonding activity he describes as wing bashing, as males compete for a female. 'You see groups of four or five clipping each other's wings, goofing around.' Like Maynard Axelson, he too has noticed the frenetic feeding behaviour of the geese in late March and April. 'They're marsh-eating machines.' What people like John and Maynard and other observers have noticed about feeding behaviour of geese at this time of year has been confirmed by studies that show that the geese reach their maximum weight in spring. Biologists believe that their lower winter weight is an adaptive strategy, that they eat less so they don't have as much to carry around. But in preparation for migration and the long nesting period ahead, the female snow geese increase their weight by as much as 50 per cent. The males gain too, but not as dramatically. At their maximum, the males weight half a kilo more than the females and are noticeably taller.

One of Sean Boyd's technicians, Saul Schneider, recalls watching a group of about 10,000 feeding on clover in a farmer's field one spring. From where he sat in his truck, about a metre away, the flock of snow geese sounded almost like a hive of bees. 'They were all making these contentment sounds.'

Hoping to watch how Saul observed the flock to determine family sizes I rode around with him as he looked for a sizeable group to survey, but it was obvious that by late March, early April, the flocks were thinning out. Some had already begun the northward flight and others – the California winterers that boosted the size of the Fraser-Skagit group in October – would not be returning because the California sector of the Wrangel population migrates north through the province of Alberta.

One day at the end of March, I sat on a mud bank above the river, downstream from a commercial fishing dock, and watched several thousand snows across a channel of river, feeding on sedges on Woodward Island. At 9:15 it was already warm, the temperature in the high teens Celsius, the mid-sixties Fahrenheit. The warm weather had promoted early growth of the marsh vegetation both here and on the

Skagit, and the geese were obviously making use of it, but from where I sat everything looked brown: the tuft of dried grass I sat on, the actual banks of the river, the river itself, in which reflections of snow geese flickered like candle flames in the current. I like being in places on earth that I can easily find on the map, and for me the mouth of a river has a particular and comfortable solidity, but on a river as large as the Fraser, it is impossible to ignore the freight the stream carries, both the obvious litter that sometimes slips out and clutters the banks, and the less visible but often more deadly constituents, such as chemical pollution. I remembered Henry Parker telling me about some foul-smelling substance that had caked his hip waders and his decoys one fall when he was out in his punt. Urban encroachment destroys wetlands, but certainly pollution must also affect the quality of the estuary. This day I found a few shell casings, a plastic milk jug, bits of white feathers caught on the weeds. Henry had said that he can get some wild shooting out here in the spring. Perhaps these feathers were left by a goose Henry's dog Tar retrieved.

On the Skagit, geese were last hunted at the end of December. Here on the Fraser during the traditional late-winter hunt, from 10 February until 10 March, hunters shot almost 300 snow geese in 1994, two-thirds of them juveniles. This is less than a third of the fall harvest of snow geese. Hunters shoot fewer birds in the spring because they are not as accessible then; most don't even return to the Fraser until the first of March. Still the hunt is controversial. The Migratory Bird Act allows for 110 days of hunting, and it is up to each jurisdiction to determine which days these will be. In the Fraser, hunting season coincides with the times the geese use the delta, which occurs in both fall and spring. Also, says Gary Grigg, head of enforcement for CWS, hunting on private lands keeps geese out of farmers' fields. He's aware, acutely aware, it seems, of the controversy surrounding the CWS decision to allow spring hunting, but in the case of snow geese, he says there's proof that the population is sufficiently healthy, and the pressure light enough that the break-up of a bonded pair does not seriously affect the population as a whole. The goose whose mate is shot would obviously see the issue differently, of course, because the loss of a mate means also the loss of a chance to reproduce that season. Widows and widowers must wait until the following spring to find partners.

As I sat watching, the sun massaging my back, I saw more pairs fly up than family groups, at least during the time I forced myself to concentrate. Though the beauty of the scene encouraged daydreaming more than structured thought, I practised disciplined observation, as I imagined a scientist would, and in ten minutes I noticed twice as many pairs as groups fly up. Thirteen pairs, six groups, and three single birds lifted into the air and flew slightly upriver to where five juvenile bald eagles were concentrating on a section of the flock. In half an hour the eagles routed the snow geese there six times.

Konrad Lorenz explained in *King Solomon's Ring* that the reason birds fear things above them is because they have an 'innate dread of the bird of prey swooping down from the heights. So everything that appears against the sky has for them something of the meaning of "bird of prey."' It's rare for an eagle to take a healthy adult bird, though: when they cruise the flock they are looking for cripples.

For the next few weeks, every time I come out to the island I must look farther and farther afield for the geese. Often I see them from the tower in the Reifel Sanctuary, far out on the fingers that strain the river before it merges with the sea. The yellowy green of the weeping willows on the island turns a darker green, red tulips replace the fading yellow of the daffs in gardens. Elderberry, sumac, wild mustard are all in white and yellow bloom, and the engines of tractors now compete with bird song. I make each trip with a mixture of hope and a slightly heavier heart, for I know the geese will be gone soon and I realize that while I have been directly and indirectly studying them for several seasons, my studies have, in a way, kept me at a distance from the geese themselves. I envy people like Maynard Axelson whose knowledge began with close contact, the kind of observation it seems I have the luxury to devote only to our cats and the birds that land on the old cherry tree in our backyard. Ironically, although I have spent more time on Westham and the Skagit this season, the geese have always been farther away than I found them before I became so interested in writing about them. I think of the old saw about how the more a person wants something, the more it eludes her.

Until I heard of the logistical difficulties humans must surpass to travel to Wrangel Island, I toyed with the idea of trying to follow the snow geese north by plane. But for now I must be content with the knowledge of migration that comes from radio collars and reports from various stations along the route. By mid-April 1994, 15,000 had been seen on the Stikine, but no one was sure if these were California geese, or the Fraser-Skagit birds beginning to appear, for all the Fraser-Skagit birds stop over on the delta of the Stikine River before moving onto Cook Inlet, near Anchorage, and then the Yukon Delta, where this story began.

In the language of the Yup'ik Eskimo people, who make up 86 per cent of the almost 20,000 people who inhabit the Yukon Delta, April is Tengmiirvik, 'the month the geese arrive.' Migrating waterfowl are so important in the culture of the Yup'ik that four months of the year have names that reflect the seasonal movement of birds through the delta. Kayangut Anutiit (May) means 'the month when eggs are laid.' Ingun (July) means 'the month of birds moulting,' and Tengun (August) means 'the month of flight [of geese].' Charles (Chuck) Hunt, a Yup'ik Eskimo who was raised in the village of Kotlik, at the north mouth of the Yukon River, and who has been working for the US Fish and Wildlife Service since 1978, describes the return of the snow geese in spring.

'They arrive [at] the delta sometime during the middle part of April as the sides of rivers, sloughs, creeks, and lakes begin to melt and form water. This is also the time when grass shoots and shoots of horsetail grass begin to grow along the sunny side of islands and river banks and sandbars.

'When the snow geese arrive in spring, the weather is usually warm, maybe thirty-five to forty degrees above zero [Fahrenheit]. At times it may be snowing wet snow, but generally ice and snow are melting. In spring we have quite a bit of daylight when the snow geese arrive. During the late part of April the sun rises around 5:30 AM and sets around 10:30 PM.

'The snow geese first arrive in spring along the Yukon River beginning just below the village of Russian Mission, Alaska. They feed along the small sand islands beginning around Marshall, Alaska, and work their way down river along the villages of Pilot Station, St. Marys, Pitka Point, and Mountain Village. Once they reach the forks of the Yukon River

Delta, they take the middle mouth and head towards the Bering Sea Coast west of Kotlik, Alaska. Here they feed for approximately two weeks. They then fly north to a location between St. Michael/Stebbins and the fish village of Pikmiktalik. Here they stay for approximately a week before they fly north to their nesting grounds on Wrangel Island.'

To appreciate the significance of the arrival of these birds, it is necessary to imagine the setting into which they fly. Picture a treeless coastal plain, foggy much of the year, threaded by meandering rivers and sloughs that pool into lakes and ponds. The Yukon is the biggest of the two rivers by far: it drains an area larger than the state of Texas and transports eighty million tonnes of silt to the Bering Sea each year. While such large deposits of alluvial soil would create productive farmland farther south, much of the Yukon-Kuskokwim Delta is underlain by permafrost. The climate is subarctic and the land is mostly tundra where sedges, grasses, and a variety of herbs and berry bushes grow. Low willow and stands of slender alder are the only wood outside of driftwood, another gift of the river and the sea.

Taken together, the Yukon and Kuskokwim form a delta about 75,000 square kilometres, most of which is protected as a wildlife refuge, the largest in the United States. A third of the delta is covered with water, making perfect habitat for migrating waterfowl, including snow geese.

The Yup'ik word for snow goose is 'kanguq,' and kanguq are a valued subsistence species in several parts of the delta, the niches where they prefer to feed, including the community of Kotlik, where Chuck Hunt grew up. Unlike emperor geese, Canadas, cackling Canadas, and brant geese, snow geese do not nest on the Yukon Delta but only stop here on their way north and south from Wrangel Island. The timing of the arrival of migrating waterfowl, at the end of a harsh subarctic winter, gives them an importance in the subsistence economy of the Yup'ik people that far outweighs the actual percentage they make up of the annual diet. For when migrating birds, including snow geese, arrive in spring they can often be the first source of fresh meat that season. According to a report by Chuck Hunt, 'some people as young as forty-five to fifty remember when, in years when seals were scarce, the spring arrival of migratory birds kept them from starving.'

Charles Hunt

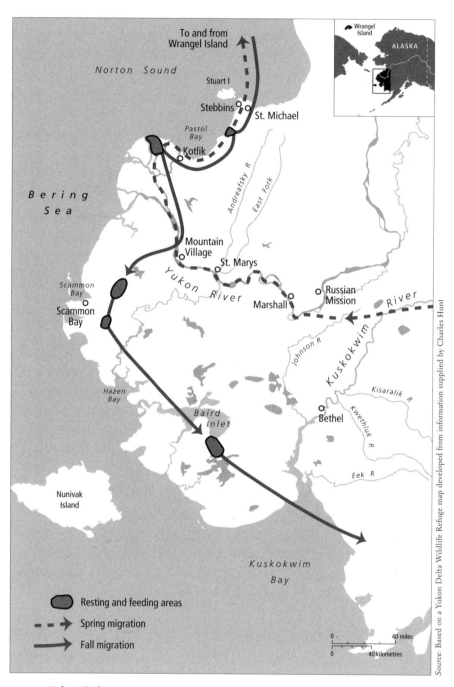

*Source:* Based on a Yukon Delta Wildlife Refuge map developed from information supplied by Charles Hunt

Yukon Delta

Because the rivers constantly change their course as stream banks erode and silt builds up, archaeologists have had a difficult time finding the materials they need to reconstruct the history of human habitation on the delta. Nevertheless, a few artifacts have been found in volcanic deposits. These indicate that humans have lived here for at least 2,000 years and that Yup'ik Eskimo life in 1600 was largely the same as it was when the Smithsonian explorer and scientist Edward William Nelson first arrived in 1877 to begin ethnological and natural history studies.

Lists Nelson made of the local fauna include snow geese, and among the hundreds of artifacts he collected were amulets and hunting pieces shaped like snow geese, masks decorated with white goose feathers, and various types of bird hunting equipment including bolas, bird spears, bird hunting arrows and snares, all of which manifest the importance of birds to the Yup'ik people.

Despite an influx of people and goods from Euro-American cultures, the Yup'ik still live off the land. This is partly because the subsistence way of life is such a deeply embedded part of the culture. As Cynthia Wentworth of the US Fish and Wildlife Service noted, in *Subsistence Waterfowl Harvest Survey, Yukon-Kuskokwim Delta*, 'The Yup'ik have always sustained themselves by living off the meat, fish, and fowl taken directly from the land and sea. The ancestors of present-day Yup'ik people undoubtedly settled on the Delta because of the very biological richness that today justifies its status as a wildlife refuge. Today the people continue to rely directly on the refuge's fish and wildlife resources for the majority of their food supply. Their culture is based on this close relationship with the natural environment.'

There is a cash economy on the delta now, and some Yup'ik people do work at service and office jobs, but even then they tend to spend their money on better hunting and fishing equipment, on fuel for their boats and their snow machines.

In the research she conducted on Yup'ik culture over several years, the Alaska anthropologist Ann Fienup-Riordan was consistently impressed by the adherence of the Yup'ik to traditional ways. 'One of the things that continually amazes me when I go back there,' she writes, in *Eskimo Essays*, 'is that people are still out there hunting, dedicated –

sometimes passionately dedicated – to continuing this way of life.'

Because of their direct dependence on the natural world, the Yup'ik have a different view of animals than Euro-Americans do. They believe that animals have immortal souls the same as humans and that animals give themselves to hunters in response to their respectful treatment of them. Fienup-Riordan writes: 'According to the world view of the Yup'ik Eskimos, human and nonhuman persons shared a number of funda-mental characteristics. First and foremost, the perishable flesh of both humans and animals was belied by the immortality of their souls. All liv-ing things participated in an endless cycle of birth and rebirth; the souls of both animals and people were part of this cycle, contingent on right thought and action by others as well as self.

'For game animals, the rebirth that followed their mortal demise was accomplished in part through the ritual consumption of their bodies by the men and women to whom they had given themselves in the chase ... During the Bladder Festival as well as the masked dances performed during Kelek, the spirits of the game were hosted and feasted as honored guests at the same time that as victims they were totally consumed.

'... Along with the belief in an essential spiritual continuity that bridged the gap between the past and the future, the Yup'ik people held that men and animals alike possessed "awareness." According to Joe Friday ... of Chevak, "We felt that all things were like us people, to the small animals like the mouse and the things like wood we liken to peo-ple as having a sense of awareness. The wood it is glad the person who is using it and the person using it is grateful to the wood for being there to be used."'

In addition to their use as food, birds, including snow geese, pro-vided the Yup'ik with feathers and skins for clothes, and with feathers and down for ceremonial purposes. The Nelson collection at the Smith-sonian includes small white geese made from beluga or walrus teeth, a parka made from emperor goose skins, and many amulets, tiny carvings of animals that were worn or carried by hunters who believed that the amulets had the power to hear, for example, walking caribou, or to see swimming geese and roaming polar bears far away. The amulets led hunters to the animals or called the animals towards the hunters and guided their bolas, arrows, lances, and harpoons. In *Inua, Spirit World of the Bering Sea Eskimo*, Fitzhugh and Kaplan explain that 'The use of white feathers of swans, gulls and geese for mask borders symbolizes the surrounding heavens and stars. Feathers of birds of prey, commonly found as helping spirits on hunting arrows and darts, are not found on masks.'

The Yup'ik view of the natural world continues in the descendants of the people Nelson met during his four years in the Bering coast region of Alaska in the late 1800s. The Yup'ik Eskimo language is widely spoken on the delta, even by children, who also learn traditional ways of interacting with the natural world by travelling with their parents and watching them hunt and fish.

'Even now, despite the introduction of western technology and education, hunting, fishing and gathering are what many Yup'ik people are most experienced in, and therefore what they tend to do best,' wrote Wentworth.

Yet spring hunting on the Pacific flyway was banned when Canada and the United States enacted the Migratory Bird Treaty in 1918. Ironically, it was the same Edward William Nelson that lived among the Yup'ik in the late 1800s who became one of the people instrumental in drawing up the act some twenty to thirty years after he left Alaska.

That an international treaty prohibits an activity the Yup'ik have engaged in for millennia has long been a sore point on the delta. The law itself was not enough to put a stop to the tradition, but scientists and wildlife managers were concerned about declining populations – not so much of snow geese, but of emperor geese and white-fronts and brant, all of which nest on the delta and had been rounded up in terrifically productive but also undeniably destructive bird drives. According to Ann Fienup-Riordan, these drives were held as late as the 1970s. 'Men, women, and children drove thousands of moulting geese and goslings across the tundra or a large lake, netted them and dispatched them by hand.'

Agencies responsible for the management of wildlife called for a voluntary halt to such practices, and the populations of threatened species such as emperor and brant geese responded to the decrease in hunting pressure, but the issue of spring hunting in the delta has remained a fiery one. The controversy demonstrates that different world views can clash dramatically over how humans should interact with the other creatures that populate the earth. For instance, many Yup'ik believe biologists harm migratory birds by hanging around their nesting areas and disturbing nesting geese by lifting eggs out of the nest to weigh and measure them. To the Yup'ik such activity violates the code of appropriate behaviour.

Fienup-Riordan explained the prescriptions and proscriptions for the culturally appropriate living of life: 'Three related ideas underlie the elaborate details of these rules: the power of a person's thought, the importance of thoughtful action in order not to injure another's mind and conversely, the danger inherent in following one's own mind.

'The qualities of personhood shared by humans and animals establish the basis for a mutual and necessary respect. Respect is understood in both positive and negative terms, including love and fear. Perhaps the most often used term is *takar* (to be shy of, respectful toward and/or intimidated by). This term, combining both admiration for and fear of the person designated, is used by juniors in reference to their relationship with elders as well as by humans in reference to their relationship with certain animals.'

The conflict of ideas is clear here as it becomes obvious that Yup'ik people, unlike scientists, did not believe there was any relationship between hunting and abundance. To the Yup'ik, it was attitude that mattered. Therefore, in their minds it would have been okay to take all the waterfowl and eggs they took as long as they did it properly.

'The admonition is to conduct one's hunting in a manner that will not offend the animals taken. Moreover, such an offense does not affect the supply of animals but simply makes them hard to find. No concept of limiting the take is apparent here, only the prohibition against disrespectful treatment of the animals that present themselves' wrote Fienup-Riordan.

Interestingly, this view appears to have been shared by aboriginal people far down the coast, who also came in contact with snow geese. The Coast Salish people William Elmendorf interviewed spoke of snow geese disappearing, or becoming less accessible and abundant, not as a result of overhunting, or of habitat depletion, or of poor weather on Wrangel Island, a place they might not even have been able to conceive of, but because a strict code of etiquette was breached by a hunter.

The Yup'ik themselves and people working on their behalf have lobbied hard for changes to the Migratory Bird Act that will take into account the views and needs of Yup'ik people who have traditionally hunted snow geese and other migratory birds when they arrive in April, a good month after the act declares that all hunting must stop.

But to convince people opposed to such an amendment that there are grounds to make it, more information was needed on just how many birds are being taken. Ironically, because spring hunting has been illegal, Yup'ik people have resisted cooperating with the agencies who normally gather harvest information. Though random surveys were done in the past, the information was outdated and no one was sure of its accuracy, and so the US Fish and Wildlife Service, in cooperation with the Yukon Delta National Wildlife Refuge, devised a unique harvest survey to be carried out by residents of the Yukon Delta themselves. Cynthia Wentworth coordinated the survey, with much help from Chuck Hunt, who works as an interpreter for the Yukon Delta Refuge, and Abraham Andrew, the head field coordinator. Thirty-one Native people have been collecting information from the residents of twenty delta villages each year since 1985.

Survey results between 1985 and 1993 reveal that about ninety pounds (or forty kilograms) of birds, or about thirty-four birds per household, were taken by residents of the Yukon-Kuskokwim Delta each year, most in spring. The harvest included an average of 1,915 snow geese each year, slightly more in the spring than in the fall, although this is, again, on average. In the spring, most are taken on the north and south coast of the delta. In the fall, most are shot on the mid-coast.

In one of his letters, Chuck Hunt described the differences in hunting techniques: 'Many years ago, before the dawn of shotguns and other firearms, Yup'ik Eskimos used bolas to hunt the snow geese. They were much easier to take than other species because they flew in large flocks. Today, shotguns are used to hunt ducks, geese, swans, and sandhill cranes. In spring, hunting is mostly done by snowmachine. Of course those who do not have snowmachines walk to wherever they hunt.'

Edward Nelson collected many types of bolas during his time on the Yukon-Kuskokwim Delta, and their use is described by Fitzhugh and Kaplan in *Inua, Spirit World of the Bering Sea Eskimo*. 'North of the Yukon, Eskimos use the bolas sling to capture low-flying ducks and geese as they crisscross between points of land and islands. Being extremely fast in flight, the birds cannot be speared or shot with arrows, but their path often takes them within sling shot of hunters concealed behind rocks or in boats.

In his 1899 reports to the Bureau of American Ethnology, Nelson provided a detailed description of hunting with bolas: 'When in search of game the bolas is worn around the hunter's head with the balls resting on his brow. When a flock of ducks, geese, or other wild fowl pass overhead, at an altitude not exceeding 40 or 50 yards, the hunter by a quick motion untwists the sling. Holding the untied ends of the cords in his right hand, he seizes the balls with the left and draws the cords so tight that they lie parallel to each other; then, as the birds come within throwing distance, he swings the balls around his head once or twice and casts them, aiming a little in front of the flock. When the balls leave the hand they are close together, the cords trail behind, and they travel so swiftly that it is difficult to follow their flight with the eye. As they begin to lose their impetus they acquire a gyrating motion, and spread apart until at their highest point they stand out to the full extent of the cords in a circle four or five feet in diameter; they seem to hang thus for a moment, then, if nothing as been encountered, turn and drop to the earth ... if a bird is struck it is enwrapped by the cords and its wings so hampered that it falls helpless.'

Bolas used on land were made from stone, ivory, or similar heavy materials that gave them greater range. But one Nelson collected from St. Lawrence Island was made of conifer wood, which meant it would float if it fell in the water.

The bolas is an historical artifact now. But Yup'ik hunters otherwise continue to use snow geese much as they did in Nelson's time, keeping the down to sew into parkas, and boiling the bird whole. The anthropologist Robert J. Wolfe reported that 'Whole plucked and gutted birds are

Hunting with bolas

commonly boiled in soups, including the heads and feet, which are considered edible parts.'

If the spring of 1995 is anything like the spring of 1993, in which Yup'ik hunters shot more snow geese than they had in the previous eight springs, some 1,500 snow geese will be shot before the flock leaves the Yukon Delta, ten to twenty of them by Chuck Hunt's brother Francis and their cousin Aloyius Unok.

Early in May, Chuck wrote to report on the progress of the season in his subarctic neighbourhood: 'Spring has been here in the Y-K Delta for the last three weeks. It has been very early this year. Most of the snow in Bethel has melted and the lakes, sloughs and rivers are now dangerous to travel on. The Kuskokwim River ice will soon be moving out. It is pretty much the same with the Yukon River. The ice thaw is now in Russian Mission. It will take about a week before the north, middle and south mouth of the Yukon River ice will become dangerous to travel.

'This year the thaw of ice and snow has been so quick that ducks, geese, swans, cranes and other migratory birds move rather quickly to their nesting areas. When this occurs, ducks and geese do not remain for a while and concentrate at melted areas such as the upper parts of the Kuskokwim and Yukon Rivers.

'For the snow geese, they too are early in their arrival. From what I have heard for this year, some of the snow geese are arriving from north of Kotlik. This means that because of the east and south winds, some of them have overshot the Yukon River and mouth of the river and are coming back south to their feeding areas along the Yukon River and west of Kotlik.

'For snow geese, whitefronts, and Canada geese, it is very difficult to identify the difference between males and females because their feather colorations are the same. So, it is then difficult to tell whether a female is more heavy in spring or not. It is also then difficult to tell if they are heavier because the eggs are growing inside them. Most likely, snow geese feed on growing grass, horsetail grass, and insects rather than plants they feed on in fall. Since I moved to Bethel in 1967 and began working for the US Fish and Wildlife Service, I have not gone hunting for snow geese in spring. But I do hunt them in the fall during late September.

A hunting party on the Yukon Delta:
Francis Hunt, Ralph Martin, and Aloysius Unok

'Since the snow geese are so close to my village, many of the hunters hunt them both in spring and fall. During the fall season it is usually my brother Francis and my cousin Aloyius Unok and I that hunt them west of Kotlik. During the third week in September before we begin our hunt we start observing and asking other hunters in Kotlik whether or not they have seen snow geese. We try not to hunt them too early. We generally wait until most of the flocks have arrived from Wrangel Island. This way, we do not disturb them early and scare them off.

'Then when we have observed heavy concentrations west of Kotlik, we set a date to go hunting. Generally, we set up camp and hunt about 2-3 days. My group uses snow goose decoys of 100-200 decoys set up between their roosting and feeding areas. We try not to hunt in their feeding or roosting areas because this may cause them to move elsewhere. In our hunts we usually average about 7-10 birds each. In fall, the weather is generally cooling off quite a bit and on several hunts we had heavy snowfall and the lakes and sloughs froze.

'When ducks and geese first arrive in spring, people generally eat the fresh birds as soon as they are brought home. Other birds are usually given to extended family members. Most hunters hunt two to three times in spring and generally stop when nesting begins. Other birds taken are usually stored whole, that is with the feathers. People now have freezers and this makes it easier to preserve meat. Prior to arrival of freezers, most people either dried or salted them for later consumption.

'In our hunts we generally do not see snow geese mating. The birds are usually too far away for us to see their activities when they are on the ground, even with binoculars, that is, to see if they are mating. Most likely, like ducks, cranes, swans and other migratory birds, they have already mated [paired] prior to leaving their wintering grounds. By the time they arrive on the Yukon and Kuskokwim Delta they are already paired and begin establishing territories and making nests. I am certain snow geese do the same. When they arrive [at] Wrangel Island, they begin establishing their territories and making their nests.'

WARM TO CHUCK HUNT is as cold as the average winter temperature on the Fraser and the Skagit Deltas. Spring in the Arctic is definitely not the tulip time it is in southwestern British Columbia and northwestern Washington. In April and early May, a linear area of water opens from the eastern Bering Sea north through the Chukchi Sea to Point Barrow as free-floating pack ice separates from more stable land-fast ice. This seemingly grudging separation of ice from ice is the beginning of spring around Wrangel Island.

# Nesting on Wrangel: The Cycle Continues

THE WORD *philopatric* stems from Greek roots, *phileo* – to love – and *pater patros* – father. Evolved for use in discussions of ecology, philopatric describes birds that have a strong urge to breed where they were raised. In the case of snow geese, though, it is not the father's land they home to so much as the mother's. For it is the female that leads the mate she chose on the wintering ground back to where she hatched and grew and fledged. A Wrangel female who spent the winter in California might have chosen a partner that grew up far to the east, on Banks Island in arctic Canada. Still her partner flies in tandem with her all 6,070 kilometres north.

It is late in May now, days are twenty hours long. The female's ovaries have begun to swell and the protein and carbohydrates she has ingested on the trip north have padded her breast with a layer of fat that will help sustain her during incubation. Considering the small window of time they have to breed successfully, the snow geese must now feel a great urgency to complete the last leg of the journey, north and west, across Bering Strait and 150 kilometres north of the tip of the Chukotka

Peninsula to Wrangel Island, a fist-shaped, treeless stretch of mountain and tundra running about 140 kilometres from east and west, and 70 kilometres from north to south.

It seems an inhospitable and far-flung place to breed, the opposite of the association of cosy with nest, but this most remote of arctic islands offers some features that colonial species such as snow geese thrive on. The many lakes, streams, and ponds on the tundra give flightless adults and their young a place to escape from arctic foxes, and the plants that fringe all these water bodies provide plenty of food for the huge colony to consume during the six weeks it takes the adults to regrow their flight feathers and the young to learn to fly for the first time. Continuous summer daylight means they can graze all day, and there are few competitors for the space and the food. Small numbers of black brant nest on Wrangel in good years, loons also breed here as do eider ducks and the beautiful Sabine's gulls and snowy owls. But the predominant avian species on Wrangel is the lesser snow goose. In the early part of the century, as many as 400,000 used to nest in various places on the island. But after 1926, when a group of Russian marine mammal scientists established the village of Ushakovskoe, Russian, Chukchi, and Eskimo residents began harvesting geese and eggs, and the snow goose population shrank. Because massive hunting and egg gathering had virtually wiped out snow geese on the Russian mainland, the Wrangel nesters were the only snow geese left in Russian territory. In all of Asia, in fact. So in 1961, the Soviet government set up an emergency preserve right around the snow goose colony. In 1976, the preserve was enlarged to encompass the entire island in order to protect the polar bears that den on the island in winter.

Though the Wrangel colony was once recognized as the largest anywhere in the world, North American ornithologists have known of it for less than forty years. The arctic explorer Vilhjalmur Stefansson tried to claim Wrangel for Canada in the mid-'20s, and reported on some of the flora and fauna that he found there, but he landed on the southern part of the island, where Mount Sovietskaya would have shielded him from the presence of the tens of thousands of geese that nested in the mountain's shadow and reared their young on the tundra that stretches to the island's northern coast. Graham Cooch only learned of the

Wrangel population in the late '50s. He had been trying to put together a model of North American goose production and came up with a quarter of a million geese in California that he couldn't account for. 'We couldn't figure out where these extra geese were coming from. We knew there was sometimes a small colony at Wales, Alaska, so I went through the Arctic bibliography and found an article by Botcharev in the *Ukraine Journal of Hunting and Fishing*. The article was written in 1932, and it told of a government hunter who lived on Wrangel Island and reported the colony of snow geese. Botcharev estimated the colony at about a quarter of a million.'

Travel to any part of the USSR was restricted during the Cold War, but during his time as arctic ornithologist for the Canadian Wildlife Service, Cooch set up a number of agreements for cooperation between North American and Soviet researchers that made access possible. Scientists now flock to the island every summer to research and catalogue, to observe and band or collar, or to carry out any number of other activities, such as sampling lake-bed sediments to determine what the climate used to be like.

I say 'flock,' but travel is considerably more difficult for humans than snow geese. The only form of transportation is helicopter, a ninety-minute flight from Cape Schmidt on the Chukotka Peninsula. And thick fog prevents flying three days out of five, so when a helicopter does arrive, it's a big event in the village of Ushakovskoe, now inhabited by about one hundred people of mixed Russian, Chukchi, and Eskimo ancestry, most of whom work for the Sapovednik, or nature reserve.

Ushakovskoe sits near a lagoon on the island's southern shore, with the mountains forming a backdrop. The highest peak is Mount Sovietskaya at 1,096 metres. The snow geese make their nests on the northern slopes of the most northerly ridge in the upper parts of the Tundra River Valley. Sean Boyd, who spent two months there in 1991, described it as a vast treeless bowl filled with 20,000 to 30,000 nests.

The land does not rise smoothly but is intersected by deep cuts and crossed by numerous streams, so that the slopes look like a series of extended hills. Because the terrain is uneven, snow accumulates more in some places than others. There are patches of ground, therefore, that thaw and dry more quickly here than in the northern flatland where, on average, snow melts seven to ten days later. So the slopes can be ideal in late spring, when the geese need large areas in which to settle and make their nests and to feed, but not much good to them when they dry out in summer. At that time, though, the vegetation around the many tundra lakes is at its peak.

Despite the advantages of a good food supply and little competition, and plenty of space to rear their young, the geese must contend with the weather. Wind so strong at times it can knock a person down. And sudden extreme changes in a single season. This happens all over the Arctic, but Wrangel is subject to even greater fluctuations because it is an island around which masses of ice float both during winter and summer. Moving ice leaves ribbons of open water that can create sudden fog, a rise in air temperature, and snow. Because the Arctic Ocean is almost never free of ice, even in the summer, temperatures on Wrangel remain cooler than in other arctic locations. Though ice may appear to be far offshore, it can suddenly return in the height of summer: the temperature will plunge and snow start to fall, even as late as July.

Everything is a trade-off. Unpredictable high arctic weather is balanced by never-ending daylight for four months of the year, which gives the geese time to lay their eggs and incubate them and raise their young all before the sun starts sinking below the horizon again, around the middle of August, after most of the goslings have begun to fly.

Long days foster quick growth of the green things the geese like to eat. Wrangel supports an amazing variety of plants, a reported 360 species, which is more than the entire Canadian Arctic: cotton sedge with its white flower head, blue-spiked lupin, wild crocus, saxifrage, arctic poppy, and ankle-high cushions of willow and juniper that, while they cannot conceal the geese from the arctic foxes that prey on them, do provide protection from the wind. Because Wrangel escaped the last glaciation, some relics of former tundra-steppe vegetation communities survive today. A small version of the woolly mammoth also survived longer on Wrangel than anywhere else. Teeth and bones found by paleontologists prove that they existed on the island up to 4,000 years ago.

Muskoxen, which were reintroduced in the 1970s, and caribou

roam the island freely, and a huge population of walrus lives just off-shore. Ninety per cent of the polar bear population shared by Russia and Alaska den on Wrangel in the winter, and a cliff on one of the capes supports the largest sea-bird colony in the eastern Siberian Arctic.

Largest. Coldest. Wrangel is a place that invites the use of superlatives, a place of extremes. Its location, the crashes and peaks of its natural cycles. Although the snow goose population stabilized after the nature preserve was established, numbers continue to rise and fall depending, primarily, on the weather during the breeding period. Sean Boyd says that there is, on average, a complete breeding failure once in every five years. In the last twenty or so years the population has gone from a low of 56,000 in 1975 to 100,000 again in 1987. In 1973, 87,000 adults arrived, but only 6,000 made nests. In 1994, an estimated 60,000 birds left the island, and perhaps 50,000 will return. Some years the geese produce many young, sometimes there is no new generation at all. In 1994 Vasily Baranyuk, who is head of snow goose research on the Wrangel Island Nature Preserve, lived at camp for six weeks alone in the snow, watching the geese struggle to nest, the few goslings that hatched to survive. He said it was the worst year for weather since the 1930s, and he has worked there for thirteen years.

Usually, by the time the geese begin to return in late spring, the scurvy grass is in perfect condition. In 1994, however, an unusually cold May prevented the foothills from thawing and therefore stifled spring growth. Frustrated, the geese were forced to waddle over crusty snow that reached up to their breast in places, searching for something to eat, the need to nest becoming more and more urgent as spring progressed. Then, in early June, there was a warm spell. The ground began to dry out, old goose couples sought out familiar nesting sites, new mothers made their initial scrapes in the gravelly soil, all the females prepared to lay their first eggs.

Imagine tens of thousands of geese whose instincts are to breed and breed quickly, marking out nest territory, copulating, hunkering down out of the wind, bracing themselves against sudden snow squalls and drops in temperature, searching for the green shoots that are beginning to push through the frigid earth. Great houcking choruses, white feathers

Snow goose pair on nest, Wrangel Island

whirling as the ganders stretch their necks straight up and squawk, lift their wings menacingly at any other gander who dares challenge his choice of nest site, tries to copulate with his mate, or share his patch of forage.

For this is a frantic time on the colony, especially in years such as 1994. Because not much ground was suitable for nesting, the fierce competition between goose pairs for nesting space intensified. In an unpublished paper on nesting ecology, Syroechkovsky reported having seen a great deal of tension between pairs during the period of settlement of the nesting territory and laying of eggs, even in years when weather was not so bad. 'Very often there are aggressive demonstrations, skirmishes and fights between males and nesting pairs. The observer can daily observe several savage fights between geese, which sometimes end in the death of one of the protagonists.' When things get too bad the geese just abandon their nests.

With less than three months before the sun touches the horizon again and the fall storms start, all the females must lay within days of each other. Syroechkovsky observed a deadline of 11 June, no matter what the weather: if nests have not been created by then, they are not created at all.

How closely one pair nests to another depends on which part of the slope the geese choose. The places where snow melts first are best because they are drier and vegetation grows sooner in the season. In areas like this there can be as many as 1,200 nests per square kilometre, while in the less desirable areas of similar proportions there may be only 600.

Most female geese breed when they are two to three years old, males at three to five years old. If a young pair tries to nest in a year when the weather is poor, they might not be able to sustain territorial challenges from older, more experienced birds. So, as Syroechkovsky found, young females may choose older males; younger males, older females. For while it is true that snow geese mate for life, and they can live for twenty years or more, hunters and other predators, accidents, and disease break up many pairs and new bonds have to be formed if an individual's line is to continue. Studies by Fred Cooke and his colleagues at La Pérouse Bay revealed that there is basically no difference in the numbers of eggs laid by females with new mates and females with mates they have bred with for several years. But clutch size does generally increase with older mothers. An older mother who loses her mate during the incubation period, however, perhaps as a result of the male unsuccessfully trying to defend his family from an arctic fox attack, has a harder time raising her brood. With no mate to stand guard, she must sit upright and alert and thus expend even more energy than other mothers on the colony.

Returning females try to use nests they've used before and often spend the days prior to laying tidying up the old nest, perhaps as a way of claiming the area. If they cannot use an old nest site they find a spot as close as possible to the place they chose the previous year.

When she cannot find a suitable place to make a nest, the female frequently lays her eggs in another goose's nest, or directly on the ground. Using someone else's nest is called nest parasitism, and the frequency with which it occurs on Wrangel Island is directly related to the shortage of space in bad weather years. Average levels of nest parasitism increase the clutch size from the mean 3.7 to between 4.2 and 4.7 eggs per nest. Geese who can't get their eggs adopted just lay them on the ground. In years when nesting conditions are bad, egg piles consisting of ten to thirty eggs are found all over the colony. Of those who do successfully nest, most lay three to four eggs, one per thirty-six hours, but their clutch may be larger because of nest parasitism.

At La Pérouse Bay researchers observed that the female who does not make a nest prefers to lay her eggs near the defended nest of another pair. Her mate stands a distance away, presumably as a way of luring the defensive nesting male away from his mate, who is beginning to incubate her natural clutch. The male is on the defensive not only because he is concerned about the survival of the eggs in his partner's nest, but because goose 'adultery' sometimes occurs, as males attempt to copulate with females that are not their partners. While her mate is busy trying to push the approaching male away, the nesting female then 'adopts' the intruding female's egg by rolling it into her nest. Adoption and extra-pair copulation are both considered to be ways of ensuring individual reproductive success.

Scientists used to think that snow geese copulated before they reached the nesting grounds, but observations at La Pérouse Bay and at

Wrangel Island in recent years have revealed that pairs commonly copulate at both sites. The following description comes from *The Snow Geese of La Pérouse Bay*, by Cooke, Rockwell, and Lank: 'At La Pérouse Bay the male usually initiates the sequence by positioning himself near his mate high in the water and with tail cocked vertically. He then carries out a series of head dips of increasing frequency. If the female is receptive, she usually adopts a bowed posture with the lower neck in the water, often accompanied by bill dipping or head dipping. The transition from pre-copulatory display to mounting is sudden. The male grasps the neck feathers and mounts the female, who crouches if she is in shallow water. Treading occurs for about 5 s[econds] and the female lifts her tail to one side as cloacal contact is made. Male snow geese have intromittent organs, which are rare in birds in general but common in waterfowl. A post-copulatory display consists of the birds rising up with vertically extended head and neck. The birds usually vocalize and flap their wings. The sequence terminates with extensive preening and bathing.'

The Wrangel snow geese behave slightly differently, going through the motions on ground instead of in the water. Both populations were observed to be otherwise similar in that they tended to copulate most just before laying and during laying. Most copulated in the early morning and no pair was seen to copulate twice in the same day, although they do it so quickly it is possible observers just did not see them doing it more than once a day.

Interestingly, the goose 'adultery' that takes place – extra-pair copulation as the scientists call it – goes forward without any ceremony at all, with none of the pre-copulatory displays observed in paired birds. 'Males often approached females on the nest, but if not, females were forced to the ground and mounted from any direction. Treading is lengthier than the usual 4-5 s during pair copulation, and the male moves his tail from side to side to slip it under the female's. In nine cases the female eventually definitely raised her tail and allowed cloacal contact ... No post-copulatory display was seen, and neither males nor females had the normal bouts of post-copulatory preening or bathing. B. Ganter observed an incubating female solicit a mating with her own mate 5 min after a forced EPC with a neighbour' (Cooke, Rockwell, and Lank 1995).

All this goes on up until the clutch is complete. During this time the female may walk off the nest to graze while the gander stands nearby to ward off intruders, but once she has laid all her eggs, the female will sit for twenty-two days, almost never leaving to find food for herself. She loses 30 per cent of her body weight, often being so weak when the eggs hatch that she does not make it with her brood and her mate to the moulting ground. The rare mother goose will starve to death right on the nest, and the gander must raise the goslings.

The strong bond between pairs was demonstrated in captive snow geese studied by researchers on Fred Cooke's La Pérouse Bay team, who observed that when pairs were separated, lone geese continued to search for their mate, even when there was no brood to watch over. The strength of the bond fosters the success of nesting in the wild, where the gander stands like a sentinel close by the nest, feeding and keeping a sharp eye out for predators.

As mentioned previously, although Siberian and hooded lemmings are the primary source of food for arctic foxes on Wrangel, they also take eggs and goslings, and in years when the lemming cycle is low, some adult snow geese, too. They destroy 30 per cent, sometimes more, of nests in a given year. Nevertheless, continuous summer daylight allows geese, who cannot see well in low light, to protect their nests quite successfully. As the brazen white fox approaches the nest, the gander stretches his neck forward, running, wings raised like twin battle shields, squawking and hissing. This is how he greets intrusions by gulls and jaegers, too. But there's another form of defence, a tactical approach. By building their nests near predatory birds such as the snowy owl, the geese can rely on the owl to scare away intruders. Since snow geese like to nest in places that dry out soonest, and this happens to be the mountain slopes, and snowy owls like to nest where the lemming populations are likely to be largest, which is also in the mountains, it works out conveniently for both species. Snowy owls do sometimes prey on snow goose nests, too, but apparently not those within their own territory, which is why snow geese, along with common eiders and black brant, often settle quite close to the nests of snowy owls.

In average years, it is early in July, three weeks after the last egg slipped down into the nest from the mother goose's body, when a slight tapping sound comes from the eggs beneath her. A series of cracks appear and begin to spread around the larger end of the eggs. Inside, the goslings are repeatedly bobbing their heads up and down, and as they do their egg tooth cracks the shell little by little. As their body shifts, the next pipping movement cracks the shell in a new place. The final egg laid will usually be last to hatch, and as is common with the last eggs of large clutches, this gosling may be weaker, the runt of the litter so to speak. In some cases, the already hatched young, who are ready to leave the nest and start foraging, prompt the mother to leave the nest before the last egg has hatched. This orphan becomes easy prey for foxes and gulls.

The goslings emerge one by one, dark yellow and wet, with greenish black legs and feet, squirming and cheeping. Within twenty-four hours, they fluff up and teeter out of the nest and begin foraging on whatever green shoots still exist on the nesting grounds, and on insects, for from now until they fledge it will be a mad rush to grow from the 100 grams they usually weigh at birth, to the 1,200 grams they will be in about six weeks, when they are ready to fly. And the mother goose must regain what she lost during incubation. The shell- and feather-littered, grazed-out slopes are no good to the snow geese now. They need the fresh green of the tundra, and so the parents lead the days-old goslings on a twenty- to sixty-kilometre trek down the Tundovaya River, sometimes swimming, sometimes walking to reach the flat, lake-dotted expanse at the northern part of the island. It's a two-week trip during which time the fractiousness of the geese during nesting is completely forgotten and the parent geese work together to ward off attacks by the persistent foxes, by gulls and jaegers and the occasional snowy owl, all of whom prey on the vulnerable young.

The snow geese that did not breed this year, including the 1,000 or so juveniles who spent their first winter on the Fraser and Skagit Deltas, have already begun to moult by the time the families arrive. They've lost the primary feathers that make the distinctive black contrast to their overall whiteness. Now certainly all of the families, and a large percentage of the non-breeding adults, cannot fly. This is when they are most vulnerable to predators and most accessible to researchers.

Snow geese families waddling down the Tundovaya River,
Wrangel Island

In 1994 it was near this moulting area, at a research camp on the northwestern part of the island, that Barbara Ganter, Don Kraege from the State of Washington, and Mike Samuel from the US Department of the Interior's Wildlife Heath Center in Madison, Wisconsin, joined Vasily Baranyuk to carry out research activities for the Pacific flyway and Fred Cooke's wildlife ecology program. Barbara had her first taste of what it means to be on a high arctic island in summer before she even reached Wrangel. Pictures she took from the helicopter showed that the sea between the Chukotka Peninsula and the island was almost all floating ice. After a day or two in Ushakovskoe, the group travelled north in a track vehicle, an old army tank that rolls over the firm gravel of the river beds to avoid damaging the tundra.

Though they could see some geese from the observation tower at the camp, they actually came into contact with them only when Vasya herded a group into the nets set up near camp. 'We could only catch the ones that were close to our camp, so we ended up only catching about 730 adults and their goslings, 360 families, a quarter to a third of what were there. We did quite well, but far less than we planned.'

They attached no markers at all to the goslings because the few they found were small for their age, and young. 'We didn't think any would make it. We know these things happen, but it's quite disappointing to see it, and also quite sad. Those birds who had decided not to nest went off to moult early. We couldn't catch them because they could fly already. So we only caught families. The goslings were still completely downy, and their parents were quite far from being able to fly. Depending on when the first frost comes, the adults might not be able to fly. We could lose some adults. It's really a very narrow window. They really reached their limits this year.'

As Sean Boyd's census later showed, some young snow geese did come out of Wrangel in 1994, but a number of these were shot by hunters or died by some other means. So that in 1995, the yearlings who gathered on the outskirts of the colony to await the start of moulting were a very small group, much smaller than the crowd of two year olds who pair bonded the winter of 1995 on the Fraser and Skagit River Deltas, and the

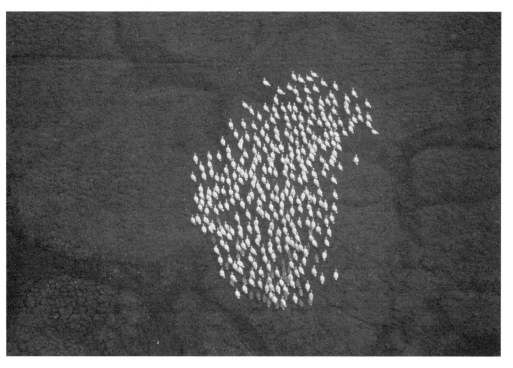

Aerial shot of snow geese on moulting grounds

rice and corn fields near Sacramento, California, and will begin to breed next year.

I had hoped to be able to end this chapter with news of favourable nesting conditions on Wrangel in 1995, a large hatch, a substantial group of healthy goslings. In fact, 1995 was as bad as 1994. The geese first began to arrive on the colony in late May when 95 per cent of it was covered with snow. On 19 June snow still clung to 50 per cent of the nesting area. Goose pairs made 4,400 nests, but only half of these were successful. And although there were fewer arctic foxes than usual hunting for eggs and goslings, a wolverine got into the colony and destroyed several hundred nests. So while the average clutch size was 4.68, larger than normal – as it almost always is during bad weather years, when nest parasitism is high – the families that left the colony had an average of less than three goslings apiece. Vasily Baranyuk estimated that 6,000 goslings left the colony this year, less than 5 per cent of the population.

In his surveys on the wintering ground, Sean Boyd found that the juvenile geese actually made up only 3 per cent of the population. Two bad weather years in a row seemed a dire situation to me, but Sean is not worried about the health of the population as a whole, for he finds that the older birds are successfully carrying it, with an overall survival rate of 85 per cent from one winter to the next. More experienced birds are less likely to fall prey to hunters and other predators, and, as has happened in the past, successive years of failure may well be followed by a big pulse of young. The story continues.

# *Afterthoughts*

I LEARNED THIS AT LEAST from my experiment ... Thus begins one of my favourite passages in literature, in Henry Thoreau's Conclusion to *Walden*. And so, with respect, I begin what must be the conclusion to at least this stage of my query with the same words. At the beginning of this project I imagined a web of views all linked by snow geese, and I pictured an actual web, like the airy silken constructions of spiders. Instead I found that the web I imagined was only surficial evidence of connection. The real networks ran underground where I found caverns full of ideas I had not known existed, gems sparkling from previously unexplored passageways.

My instincts were right in one respect, however: snow geese do demonstrate how distant places and people are tied to one another. Any natural history of a wild bird or mammal or fish must include humans because, of course, humans are part of the natural world; the action of each individual in the chain of snow goose observers affects all the other links in the chain, as well as the geese. Yet today, with the great remove at which many humans live from nature, that idea has to be hawked by conservationists and scientists such as Edward O. Wilson and Stephen Kellert who, in trying to convince decision makers of the importance of maintaining the diversity of life on earth, go to extravagant lengths to prove affiliation. Their work reflects a general uneasiness, an insecurity about the human place in the big picture.

The people who seem to have fewest doubts about their relationship with the natural world are the Yup'ik who developed a code of behaviour meant to perpetuate the animals and plants they traditionally depended upon. The confidence they have in their system prompts a romantic longing in some people to be like them, though we cannot erase from our brains the impressions that have made us who we are. And there are other codes, the code of the sportsman handed down from European ancestors; the code of the naturalist who agrees to take nothing from the wilderness but pictures, leave nothing but footprints. As for the Yup'ik, the natural world they know at the end of the second millennium is expanding as they travel by plane or by television to places they may not even have imagined a century ago.

The snow geese attract us because they travel in such big flocks, create such a dramatically beautiful spectacle, but when focusing on the flock, we lose sight of the individual. Individual snow geese persisted in looking weird to me, out of sync. Ironically, however, it is as individuals that we humans must work out our fit in the world, as individuals dealing with other individuals, whose spot in the puzzle is as uniquely shaped as our own. Information is crucial, and scientists can provide much of that; but so can observers who come by their knowledge through their intimate connection with the seasons, and this includes hunters, Native and non-Native, and farmers; and so can artists, who show us new ways to look.

Consider the various specialists in our species as parts of the human body, the scientists, perhaps, being the head; the artists, the heart; the hunters and farmers the hands; the Indian and Eskimo, the spirit. A full relationship with other species inhabiting the planet obviously requires participation by the whole.

Those who strive for an objective view forget that objects haven't the ability to see. The challenge is to broaden one's own view so that it takes into account the views of others. Instinctively understanding the importance of this, someone long ago thought of the town hall meeting, the public forum meant to give people an opportunity to hear one another's thoughts. It's still a good idea. But you can't hear if you don't listen, and why listen if you think you already have all the answers?

I learned this at least from my experiment, that, as Maynard Axelson told me, 'It's not the fish, it's the fishing.' Answers are not so important as asking questions. Using one's own life to become more aware of life in general. I recall the man whose example inspired my personal search for greater awareness, the late wildlife biologist and bird artist Glen Smith. I was visiting Glen and his wife at their Salt Spring Island home one summer, sitting out on the deck with Glen, talking, when he paused to watch the progress of a falling seed. It landed near us and he picked it up to show me how the point at its end helped it to lodge somewhere firmly, so that it would have a better chance to germinate, to grow and produce more seeds. The view of a biologist, all about life, continuance.

On the 31st of May I made another visit to Westham Island, heavy with lush green plants, ploughed fields sprouting new growth. I wanted to take a picture of John Ireland in his goose hat, and he obliged, suggesting he pose in the pen behind one of the outbuildings, where he had a snow goose in captivity. A bird someone had found on a beach at the far western edge of the city, below the University of British Columbia, and brought to him. The bird's left wing was so badly damaged it had to be amputated. Still the snow goose resisted John's attempts to gather it into his arms, and dashed here and there, squawking. He caught it before long, though, and after I snapped the picture he let me touch it. The first live snow goose I had touched. I stroked its curved head, spoke to it as I would a puppy. It tolerated my attentions for a minute, then curled its neck around in frustration, maybe wondering how it had ended up in this situation, no longer able to fly up to the Arctic with its mate, but condemned now to spend its days in a pen, fawned over by humans who knew it little or not at all.

# Bibliography

Bateman, Robert, and Rick Archbold. 1980. *An Artist in Nature*. New York: Random House

Bateman, Robert, and Ramsay Derry. 1981. *The Art of Robert Bateman*. New York: Viking Press

—. 1985. *The World of Robert Bateman*. Markham, ON: Viking/Penguin

Bellrose, Frank C. 1980. *Ducks, Geese and Swans of North America*. Harrisburg, PA: Stackpole Books

Bolles, Edmund Blair. 1991. *A Second Way of Knowing: The Riddle of Human Perception*. New York: Prentice-Hall

British Columbia Ministry of Environment. 1980. *Preliminary Snow Goose and White-fronted Goose Management Plan for BC*. Victoria: Government of British Columbia

Buckhout, Robert. 1974. Eyewitness Testimony. *Scientific American*, December, 205-13

Caldara, Anna Maria. 1992. *How Birds Fly*. Toronto: Doubleday

Caras, Roger. 1985. *The Endless Migrations*. New York: Dutton

Cooch, E.G., R.L. Jeffries, R.F. Rockwell, and Fred Cooke. 1993. Environmental Change and the Cost of Philopatry: An Example in the Lesser Snow Goose. *Oecologia* 93:128-38

Cooch, E.G., D.B. Lank, R.F. Rockwell, and F. Cooke. 1989. Long Term Decline in Fecundity in a Snow Goose Population: Evidence for Density Dependence? *Journal of Animal Ecology* 58:711-26

—. 1991. Long Term Decline in Body Size in a Snow Goose Population: Evidence of Environmental Degradation. *Journal of Animal Ecology* 60:483-96

Cooch, E.G., L.C. Newell, R.F. Rockwell, and F. Cooke. 1990. Queen's University Tundra Biology Station Lesser Snow Goose Project Summer Guide. Unpublished

Cooke, Fred. 1980. Genetic Studies of Birds: The Goose with the Blue Genes. Plenary Lecture. Dept. of Biology, Queen's University, Kingston, ON

—. 1983. Goose Mates. *Natural History* 1:37-42

—. 1987. Avian Genetics. In *Lesser Snow Goose: A Long Term Population Study*, 407-32. London: Academic Press

Cooke, Fred, M.A. Bousfield, and A. Sadura. 1981. Mate Change and Reproductive Success in the Lesser Snow Goose. *Condor* 83:322-7

Cooke, Fred, Robert Rockwell, and David Lank. 1995. *The Snow Geese of La Pérouse Bay: A Study of Natural Selection in the Wild*. New York: Oxford University Press

Cooke, Fred, and Daniel S. Sulzbach. 1978. Mortality, Emigration and Separation of Mated Snow Geese. *Journal of Wildlife Management* 42(2):271-80

Collins, June McCormick. 1974. *The Valley of the Spirits*. Seattle: University of Washington Press

Dobb, Edward. 1995. Without Earth There Is No Heaven. *Harper's Magazine*, February

Elmendorf, William. 1993. *Twana Narratives*. Seattle: University of Washington Press; Vancouver: UBC Press

Fienup-Riordan, Ann. 1990. *Eskimo Essays*. New Brunswick, NJ: Rutgers University Press

Finney, George, and Fred Cooke. 1978. Reproductive Habits in the Snow Goose: The Influence of Female Age. *Condor* 80:147-58

Fitzhugh, William W., and Susan A. Kaplan. 1982. *Inua, Spirit World of the Bering Sea Eskimo*. Washington: Smithsonian Institution Press

Francis, Charles M., Miriam H. Richards, Fred Cooke, and Robert Rockwell. 1992. Long Term Changes in Survival Rates of Lesser Snow Geese. *Ecology* 73(4):1,346-62

—. 1992. Changes in Survival Rates of Lesser Snow Geese with Age and Breeding Status. *The Auk* 109:731-4

Franck, Frederick. 1973. *The Zen of Seeing: Seeing/Drawing as Meditation*. Toronto and New York: Random House

Freethy, Ron. 1982. *How Birds Work*. Poole, Dorset: Blandford Books

Gallico, Paul. 1941. *The Snow Goose*. New York: A.A. Knopf

Ganter, Barbara. 1994. Wrangel Island: Biology, Conservation Problems. Unpublished

Grass, Al. 1995. A Tale of Cattails and Bulrushes. *Marshnotes* (Spring):11

Harper, J. Russell. 1971. *Paul Kane's Frontier*. Toronto: University of Toronto Press

Hills, Rust. 1987. *Writing in General and the Short Story in Particular*. Boston: Houghton Mifflin

Kellert, Stephen J., and Edward O. Wilson. 1993. *The Biophilia Hypothesis*. Washington, DC: Island Press

Johnsgard, Paul A. 1965. *Handbook of Waterfowl Behavior*. Ithaca, NY: Cornell University Press

—. 1968. *Waterfowl: Their Behaviour and Natural History*. Lincoln: University of Nebraska Press

—. 1974. *Song of the North Wind*. New York: Anchor Press/Doubleday

—. 1978. *Ducks, Geese and Swans of the World*. Lincoln: University of Nebraska Press

Kohl, Judith, and Herbert Kohl. 1977. *Notes from the View from the Oak*. New York: Sierra Club Books, Charles Scribner and Sons

Lank, David B., Marjorie Bousfield, Fred Cooke, and Robert Rockwell. 1991. Why Do Snow Geese Adopt Eggs? *International Society for Behaviourial Ecology* 2:181-7

Lank, David B., Pierre Mineau, Robert F. Rockwell, and Fred Cooke. 1989. Intraspecific Nest Parasitism and Extra-pair Copulation in Lesser Snow Geese. *Animal Behaviour* 37:74-89

Leach, Barry. 1982. *Waterfowl on a Pacific Estuary*. Victoria: British Columbia Provincial Museum

Leopold, Aldo. 1966. *A Sand County Almanac*. New York: Oxford University Press

Lorenz, Konrad. 1952. *King Solomon's Ring*. London: Methuen

—. 1970. *Studies in Animal and Human Behaviour*. London: Methuen

Macdonald, Jake. 1993. The Reifel Migratory Bird Sanctuary Is the Product of an Unlikely Alliance. *Conservator* 14(1):3-5

*Macgregor's 1994-95 Skagit County Visitors and Newcomers Guide*. 1994. Mount Vernon, WA: Macgregor Publishing

Martin, K., F.G. Cooch, R.F. Rockwell, and F. Cooke. 1985. Reproductive Performance in Lesser Snow Geese: Are Two Parents Essential? *Behaviourial Ecology and Sociobiology* 17:257-63.

McKelvey, Rick. 1989. Lesser Snow Goose. *Hinterland Who's Who*. Ottawa: Canadian Wildlife Service. Environment Canada

Nelson, Edward W. 1899. *The Eskimo about Bering Strait*. Bureau of American Ethnology Annual Report, Vol. 1. Washington, DC: Smithsonian Institution

O'Casey, Sean. 1973. *Rose and Crown*. London: Pan Books

Percy, Walker. 1989. The Fateful Rift: The San Andreas Fault in the Modern Mind. 18th Annual Jefferson Lecture, 3 May. Reprinted as The Divided Creature in *Wilson Quarterly* (Summer 1989):77-87

Peterson, Roger Tory. 1990. *A Field Guide to Western Birds*. New York: Houghton Mifflin

Robbins, Chandler S., Bertel Bruun, and Herbert S. Zimm. 1983. *A Guide to Field Identification: Birds of North America*. New York: Golden Press

Sage, Bryan. 1986. *The Arctic and Its Wildlife*. New York: Facts on File Publications

Siebert, Charles. 1993. The Artifice of the Natural. *Harper's Magazine*, February, 43-51

Smythe, R.H. 1975. *Vision in the Animal World*. London: Macmillan

Syroechkovsky, E.V., and Fred Cooke. N.d. A Comparison of the Nesting Ecology of the Lesser Snow Geese of La Pérouse Bay, Manitoba, and of Wrangel Island, Chukotka, USSR. Unpublished

Syroechkovsky, E.V., F. Cooke, and W.J.L. Sladen. N.d. Population Structure of the Lesser Snow Geese of Wrangel Island, Russia. Unpublished

Thoreau, Henry David. 1942. *Walden*. New York: Signet Books, New American Library

Thornton, H.R. 1931. *Among the Eskimos of Wales, Alaska, 1890-93*. Baltimore: Johns Hopkins University Press

Walls, G.L. 1942. *The Vertebrate Eye*. Bloomfield Hills, MI: Cranbrook Institute of Sciences

Wentworth, Cynthia. 1994. *Subsistence Waterfowl Harvest Survey, Yukon-Kuskokwim Delta*. Anchorage and Bethel, AK: US Fish and Wildlife Service/Yukon Delta National Wildlife Refuge

Wolfe, Robert J., Amy W. Paige, and Cheryl L. Scott. 1990. *The Subsistence Harvest of Migratory Birds in Alaska*. Technical Paper No. 197. Juneau: Alaska Department of Fish and Game

# Index

Printed and bound in Canada by Friesens
Cartographer: Eric Leinberger
Copy-editor: Barbara Tessman
Designer: George Vaitkunas
Proofreader: Gail Copeland